T0341203

Advanced Studies in Multi-Criteria Decision Making

Series in Operations Research

Series Editors:
Malgorzata Sterna, Marco Laumanns

About the Series

The CRC Press Series in Operations Research encompasses books that contribute to the methodology of Operations Research and applying advanced analytical methods to help make better decisions.

The scope of the series is wide, including innovative applications of Operations Research which describe novel ways to solve real-world problems, with examples drawn from industrial, computing, engineering, and business applications. The series explores the latest developments in Theory and Methodology, and presents original research results contributing to the methodology of Operations Research, and to its theoretical foundations.

Featuring a broad range of reference works, textbooks and handbooks, the books in this Series will appeal not only to researchers, practitioners and students in the mathematical community, but also to engineers, physicists, and computer scientists. The inclusion of real examples and applications is highly encouraged in all of our books.

Rational Queueing
Refael Hassin

Introduction to Theory of Optimization in Euclidean Space
Samia Challal

Handbook of The Shapley Value
Encarnación Algaba, Vito Fragnelli and Joaquín Sánchez-Soriano

Advanced Studies in Multi-Criteria Decision Making
Edited by Sarah Ben Amor, Adiel Teixeira de Almeida,
João Luís de Miranda, and Emel Aktas

For more information about this series please visit:
https://www.crcpress.com/Chapman--HallCRC-Series-in-Operations-Research/book-series/CRCOPSRES

Advanced Studies in Multi-Criteria Decision Making

Edited by

Sarah Ben Amor
Telfer School of Management
University of Ottawa

Adiel Teixeira de Almeida
Universidade Federal de Pernambuco

João Luís de Miranda
Instituto Politécnico de Portalegre
CERENA, Instituto Superior Técnico

Emel Aktas
Cranfield School of Management

CRC Press is an imprint of the
Taylor & Francis Group, an informa business
A CHAPMAN & HALL BOOK

CRC Press
Taylor & Francis Group
52 Vanderbilt Avenue,
New York, NY 10017

International Standard Book Number-13: 978-1-138-74388-5 (Hardback)

Visit the Taylor & Francis Web site at
http://www.taylorandfrancis.com

and the CRC Press Web site at
http://www.crcpress.com

Contents

Foreword

Multiple criteria decision making (MCDM) is an effective approach to structuring a complex problem and exploring meaningful courses of actions to converge to good solutions that balance various concerns of decision makers. The need to do this is becoming more crucial as the challenges the planet and the societies are facing get more complex and the consequences get more grave. There seems to be an agreement among scientists that climate change has passed certain thresholds and some of the potential disasterous effects to the planet are now irreversible. In addition to climate change, racism, access to healthcare, lack of education, unemployment, immigration, and poverty are some of the major problems faced by masses in the twenty-first century. Many of those who are in positions to make changes, however, seem to overlook these major problems. We seem to be far from the necessary vision and collaboration to start making progress on these urgent issues. The efforts of many nongovernmental organizations to attract societies' attention to some of these problems are commendable but not sufficient to reverse the negative effects. This is where I believe MCDM scholars can make a difference. Studying such complex problems that have the potential to ruin many lives of future generations may make a positive impact. We have the capability of structuring, exploring, and demonstrating the consequences of various decisions (especially the business-as-usual scenarios). Disseminating these results not only in scholarly publications but also in mass media can increase the awarenesses of the societies and may help initiate major changes in the right direction.

I personally know the editors and many of the authors of this book. They have been making important methodological and practical contributions to MCDM. I have served the International Society on MCDM for many years in different capacities including as president of the society for 4 years. During my tenure at these positions, I have known and collaborated with many MCDM scholars including the editors and authors of this book. Many of the works published in this book were presented at the 24th International Conference on MCDM held in Ottawa, Canada in July 2017. Sarah Ben Amor, the lead editor of this book, and her colleagues organized the conference. The theme of the conference was "Creating a Sustainable Society," fitting well with the concerns I mentioned above. The conference was memorable both scientifically and socially. There were plenary talks on climate change and sustainable healthcare, as well as regular talks on complex societal problems. This book is a good reflection of the rich content of the conference and it is an important step in the direction our field should grow in order to make important contributions to complex environmental and socio-economical problems. Some of the topics the book covers are major trends in today's world from an MCDM perspective, and applications in the areas of healthcare, sustainable planning, telecommunication, agriculture, and planning under uncerainty.

The MCDM community is large and very international; The International Society on MCDM currently has over 2700 members from about 100 different countries. Conferences once every two years typically attract 300–500 scholars from about 40 different countries. The MCDM summer schools held every two years bring some of the best instructors to interact with about 50 PhD students coming from all over the world. I would like to see young researchers follow the lead of this book and collaborate more with experienced researchers as well as those from different disciplines to address the challenging problems that are threatening our planet and societies. After all, MCDM scholars are among the best equipped researchers to make differences in these urgent issues.

Murat Köksalan

President, International Society on MCDM, 2015–2019

Ann Arbor, Michigan

Preface

THE BOOK *ADVANCED STUDIES IN MULTI-CRITERIA DECISION MAKING* presents a state-of-the-art, international collection of contributions about recent Multi-Criteria Decision Aiding/Making (MCDA/M) developments. Given that Decision Sciences are recognized today as indispensable for confronting the major societal challenges in science and technology, the book addresses a set of topics in which MCDA/M is crucial in today's digital reality. Without the proper MCDA/M tools, the necessary developments and innovative research would be impeded, making it harder to answer growing global problems in areas such as climate change, energy and transportation, healthcare and social sustainability—with all their diverse repercussions within the national and local contexts.

Most of the studies in this volume are developed within the international cooperation framework for R&DI projects. The contributing authors come from many different countries, and the topics of the chapters originated in MCDM-2017 (http://sites.telfer.uottawa.ca/mcdm2017/), the international conference of the prestigious *International Society on Multiple Criteria Decision Making* that brought many of them together. The conference was held in Ottawa (Ontario, Canada) in July 2017, which was also Canada's 150th anniversary.

In Chapter 1, H. Wallenius and J. Wallenius provide an overview of the mega-trends that are transforming the world, with a focus on technology transformations that are of interest from an MCDM perspective. They discuss the role that MCDM could play in these mega-trends, as well as how mega-trends have been changing MCDM.

In Chapter 2, Clímaco and Craveirinha highlight how the rapid evolution of new telecommunication technologies and services has given rise to a growing interest in applying multi-criteria evaluation approaches in a wide variety of decision-making processes involved in network planning and design. The authors provide an overview of contributions, critical evolutions, challenges and future trends concerning the applications of MCDA/M in telecommunication network planning and design.

In Chapter 3, Norese introduces SISTI, a methodological multicriteria modelling approach to structure a new and complex problem and to elaborate and validate a new model when decision makers do not exist, cannot participate or do not want to be involved in the decision-aiding process. This approach is especially effective for new practitioners to help them understand what a "good" model is and how the robustness of their conclusions can be improved.

In Chapter 4, Polyashuk focuses on multiple-criteria models for decision-making situations with a complex set of criteria. More specifically, she explores different ways to treat quantitative (tangible) and qualitative (intangible) criteria in a model aiming at approximating decision maker's preferences in an efficient and unbiased manner.

In Chapter 5, Dopazo and Martínez-Cespedes present methods and algorithms for smart-city rankings. They propose a two-stage approach to address the group-ranking problem in the smart city context. Their approach is based on deriving the priority vectors of cities from outranking matrices that collect relevant information from input data. The application of the proposed methods is illustrated using the data provided by the IESE Cities in the Motion Index 2016 (CIMI 2016) report. Their approach provides a theoretical framework for studying the problem, efficient computational methods to solve it and some performance measures.

In Chapter 6, Aguirre and Manyoma examine agricultural supply-chains prioritization for the development of areas affected by the military conflict in Colombia. Prioritization is necessary in national and international organizations to effectively direct their resources toward the development of the incipient agro-chains of the region. Using MCDA, the authors provide a ranking of the agro-chains that best represent this region of the country.

In Chapter 7, Miranda, Nagy and Casquilho examine decision-making and robust optimization for medicines shortages in pharmaceutical supply chains. The main topics of the COST Action "Medicines Shortages" (CA15105) are introduced, and they discuss how MCDM tools can be used to address the suppliers-selection problem and to curb shortages. A case-study that involves a supplier bid is analyzed using four different MCDM methods and resulting in the selection of one of the bidder-supplier companies.

In Chapter 8, Brison, Delbaere, and Pirlot adapted spatial decision models to address the following question: is it possible to rank chocolates with different degrees of fat bloom (i.e., a white-grayish layer or white spots on their surface due to fat recrystallization) without an expert panel? More specifically, models that were initially developed to help decision-makers express their preferences over maps representing the state of a given territory at different times were applied to rank chocolates.

In Chapter 9, Skulimowski proposes a model in which anticipatory decision-making principles are integrated with multicriteria sustainable planning. The model is applied on a real-life case-study to analyze the planning of the future operation of an innovative digital knowledge platform with respect to multiple criteria related to financial sustainability, technological excellence and social benefits. This platform has been developed within an ongoing EU Horizon-2020 research project (cf. www.moving-project.eu).

In Chapter 10, Kandakoglu and Ben Amor propose a robust multiple-criteria approach to select a Course Of Action (COA) in a military operation-planning process. The approach is based on the SMAA-PROMETHEE method that performs Monte-Carlo simulations and runs PROMETHEE to investigate the robustness of COA rankings when input parameters are uncertain or incomplete. The main advantage of this approach is its ability to articulate to the commander why one COA is preferable to another by exploring the input-parameter space that assigns a given COA to a certain rank.

In Chapter 11, Kilic and Kabak analyze the relationship between human development and competitiveness using the combined approach of Data Envelopment Analysis and cluster analysis. Using this approach, 56 countries are evaluated and ranked for the years 2010–2017 based on the data of the Global Competitiveness Index and Human Development Index.

With these contributions, the book presents an updated picture of the landscape of Decision Sciences, their current research topics, their interaction with other sciences, their useful collaborations with industry and services, as well as recent or ongoing international challenges.

The chapters of this volume, with relevant contributions about the application of Decision Sciences and their tools, are of interest to a broad spectrum of readers who wish to gain a fresh insight into the MCDA/M state-of-the-art, including decision-makers, managers, researchers, and MSc/PhD students.

At last, we would like to express our appreciation and gratitude to all the authors for their quality contributions, as well as we very much thank the reviewers too for their time and valuable inputs.

Sarah Ben Amor

Adiel Teixeira de Almeida

João Luís de Miranda

Emel Aktas

Editors

Sarah Ben Amor holds an MSc and a PhD in Business Administration, specializing in operations and decision support. Her research is focused on multi-criteria decision making. It looks mainly at uncertainty modeling, information imperfections, and how they are treated in multi-criteria decision analysis. Her expertise in model building and uncertainties associated with multi-criteria analysis has benefited various R&D projects for Defence R&D Canada–Valcartier, particularly with regard to risk analysis. She also has numerous applications in different fields such as finance, innovation, and healthcare systems.

Adiel Teixeira de Almeida is Professor of Management Engineering at Universidade Federal de Pernambuco and founding coordinator of the Center for Decision Systems and Information Development (CDSID). He holds a PhD in management engineering from the University of Birmingham, UK. His main interests are in decision making related to multiple objectives and group decision problems, which includes methodological issues and applications. Also, he has been working as a consultant and with R&D projects for private and public organizations, where he has applied decision models in many contexts, such as risk analysis, reliability and maintenance, project portfolio, R&D project portfolio, project management, strategic modeling, outsourcing, information systems, supply chain, and water management. He has authored or coauthored more than 120 scientific papers in reviewed journals related to a variety of topics such as Operational Research, Group Decision and Negotiation (GDN), Decision Systems, MCDM/A (Multi-Criteria Decision Making and Aid), Risk, Reliability, Maintenance, Safety, Quality, and Water Resources and serves on the editorial board of some scholarly journals, including *GDN Journal*, *IMA Journal of Management Mathematics*, *International Journal of Decision Support System Technology*, and *EURO Journal on*

Decision Processes. He has been an active member of the main societies related to Operational Research, Group Decision, MCDM/A topics. Currently, he serves the GDN Section of INFORMS as Vice-President and served, until 2019, the council of the MCDM Section of INFORMS and the Executive Committee of the International Society on Multiple Criteria Decision Making. He is an Associate Research Fellow of the Institute of Mathematics and its Applications (FIMA). He also received in 2017 the INFORMS GDN Section Award.

João Luís de Miranda is adjunct professor (tenured) at College of Technology and Management (Portalegre Polytechnics Institute, Portugal) and researcher in Optimization methods and Process Systems Engineering (PSE) at CERENA (Instituto Superior Técnico, Lisboa). He has been teaching for about two decades in the Mathematics group (mainly Calculus, Operations Research, Numerical Analysis, Quantitative Methods, Multivariate Analysis), and he is author and editor of several publications in Optimization, PSE, and Education subjects in Engineering and OR/MS contexts. He is also interested in strength the referred subjects through international cooperation in multidisciplinary frameworks.

Emel Aktas holds a Chair of Supply Chain Analytics and Professor at Cranfield School of Management. She specializes in mathematical modeling, simulation, decision support systems, and statistical analysis to address supply chains problems, specifically in transport, retail, and manufacturing sectors. Her recent research focuses on food supply chain management, with one project (SAFE-Q) on minimizing the waste in food supply chains and another (U-TURN) on logistics collaboration practices for distribution of food in the cities. Her work has appeared in *European Journal of Operational Research*, *Interfaces*, *International Journal of Production Economics*, and *Computers and Human Behaviour.*

Contributors

Eduar Aguirre
Area of Logistic Engineering
Universidad del Valle
Yumbo, Colombia

Sarah Ben Amor
Telfer School of Management
University of Ottawa
Ottawa, Ontario, Canada

Valérie Brison
Faculty of Engineering
University of Mons
Mons, Belgium

Miguel Casquilho
Department of Chemical Engineering
Instituto Superior Técnico
and
CERENA "Centro de Recursos Naturais e Ambiente"
Instituto Superior Técnico
Universidade de Lisboa
Lisboa, Portugal

João Clímaco
Institute for Systems Engineering and Computers at Coimbra
University of Coimbra
Coimbra, Portugal

José Craveirinha
Institute for Systems Engineering and Computers at Coimbra
University of Coimbra
Coimbra, Portugal

Claudia Delbaere
Cacaolab bvba
Evergem, Belgium

and

Faculty of Bioscience Engineering
Department of Food Technology, Safety and Health
Laboratory of Food Technology and Engineering
Ghent University
Ghent, Belgium

Koen Dewettinck
Faculty of Bioscience Engineering
Department of Food Technology Safety and Health
Laboratory of Food Technology and Engineering
Ghent University
Ghent, Belgium

Esther Dopazo
Computer Science School
Universidad Politénica de Madrid
Madrid, Spain

Özgür Kabak
Industrial Engineering Department
Istanbul Technical University
Istanbul, Turkey

Ahmet Kandakoglu
Telfer School of Management
University of Ottawa
Ottawa, Ontario, Canada

Hakan Kılıç
School of Sciences & Engineering
Koç University
Istanbul, Turkey

Pablo Manyoma
School of Industrial Engineering
Universidad del Valle
Cali, Colombia

María L. Martínez-Céspedes
Computer Science School
Universidad Politénica de Madrid
Madrid, Spain

João Luís de Miranda
Instituto Politécnico de Portalegre
Portalegre, Portugal

and

CERENA, "Centro de Recursos Naturais e Ambiente"
Instituto Superior Técnico
Universidade de Lisboa
Lisboa, Portugal

Mariana Nagy
Faculty of Exact Sciences
"Aurel Vlaicu" University of Arad
Arad, Romania

Maria Franca Norese
Politecnico di Torino
Department of Management and Production Engineering
Turin, Italy

Marina V. Polyashuk
Department of Mathematics
Northeastern Illinois University
Chicago, Illinois

Marc Pirlot
Faculty of Engineering
University of Mons
Mons, Belgium

Andrzej M.J. Skulimowski
Decision Science Laboratory
Department of Automatics and Robotics
AGH University of Science and Technology
and
International Centre for Decision Sciences and Forecasting
Progress & Business Foundation
Kraków, Poland

Hannele Wallenius
Aalto University School of Business
Aalto University
Helsinki, Finland

Jyrki Wallenius
Aalto University School of Business
Aalto University
Helsinki, Finland

CHAPTER 1

Implications of World Mega Trends for MCDM Research

Hannele Wallenius and Jyrki Wallenius

CONTENTS

1.1 INTRODUCTION

Digital technology is making rapid advances. The implications for people, companies, and societies are pervasive. It is difficult to foresee all the changes these developments will cause. Understandably, most individuals, many businesses and government leaders are not aware of, let alone prepared for the future changes. According to Brechbuhl from Dartmouth College, this ignorance was the driver behind the recent report, *Deep Shift: Technology Tipping Points and Societal Impact*, of the World Economic Forum.

The envisioned changes will bring about (1) digital connectivity, independent of time and place, and (2) tools for quickly analyzing vast amounts of digital data. In the World Economic Forum's report, the changes are grouped into six "mega-trends." We borrow freely from the report.

1. The Internet—world's access to the Internet will continue improving; people's interaction with it will become more ubiquitous
2. Further enhancements in computing power, communications technologies, and data storage, and the ability to interface with digital technology, anytime using multiple devices
3. The "Internet of Things"
4. Big data and Artificial Intelligence (AI)—the ability to access and analyze huge amounts of data; coupled with the "ability" of computers to make decisions based on this data
5. The sharing (or platform) economy and distributed trust (based on, for example, the block chain technology)
6. 3D-printing

These trends will greatly impact our lives, businesses, and governments—even universities—all around the world. As the World Economic Forum's Report astutely observes, our lives are increasingly being driven and enabled by software. The envisioned changes will be so profound and rapid that large segments of societies have difficulty in keeping up with the developments as users of technology.

The potential of the digital technology is huge, both in enhancing traditional industrial processes (robotics), and even more importantly in generating novel digital services. Many aspects of health care are also benefiting tremendously from new technologies. The digital revolution has begun, although decades (centuries) are needed for its full potential to be realized. One interesting cause of the Internet and social media (which totalitarian governments try to control) is the increased transparency of societies, which helps to improve democracy.

Besides technology mega-trends, there are other highly important mega-trends. These mega-trends, unlike technology mega-trends, are generally perceived as challenges or threats to humankind. Some of them are discussed in PwCForesight#megatrends and by the World Economic Forum:

1. Demographic and social change taking place in many countries (aging populations, decreasing fertility, urbanization, refugee problem)
2. Increasing world population: growing need for food, clean water, and cheap energy
3. Climate change, concern for environment

The mega-trends, whether technology related or non-technology related, pose real concerns, challenges, or even threats to humankind. Most certainly, all of these mega-trends force governments and businesses to operate more efficiently under resource scarcity. Regarding technology mega-trends, privacy issues and security issues are not easy to solve, and today's societies are grappling with them. Moreover, with robots/AI "outsmarting" many individuals (with time, perhaps most individuals), what do most people do in year 2118? Brechbuhl asks the good question, "What will happen to the sense of worth, place, and contribution to society that human beings have derived from work throughout much of recorded history?" To make matters worse, who guarantees that the AI-driven robots are (programmed to be) friendly toward humankind?[1]

We choose technology mega-trends 1, 4, and 5, and non-technology mega-trend 3 from the World Economic Forum's list, for a closer look. What role can multi-criteria decision making (MCDM) play in them? How can MCDM help? What MCDM concepts will be useful? Recall that our lives are increasingly being driven and enabled by software. We think that it is a good starting point that many MCDM scholars can write their own software. Hence, we should be able to provide tools, software, and ideas to capitalize on rising opportunities and tackle problems resulting from the world's mega-trends.

1.2 INTERNET SEARCHES

E-commerce is continuing to transform commerce. To an increasing extent people make purchases online. Surprisingly (to us), besides travel and leisure industries, the clothing or fashion industry is almost driving the change. Typically when people buy online, they use some search engines, such as Google. It is not uncommon that the cheapest products

[1] Physicist Stephen Hawking (1943–2018), among other famous people, is concerned about this.

or services emerge on top of the list. A typical example is flight tickets between two cities. Incidentally, this apparently is forcing airlines to adopt the strategy originally followed by low-cost airlines of charging extra for better seats, meals, baggage, etc. One problem is that the search engines are not good enough in differentiating among offers (what they actually contain and how much customers value if a bag or meal is included in the price). MCDM scholars could develop better search engines! Search engines, which would not only be based on price, but other attributes as well. Keyword searches have their limitations.

Because of the abundance of offerings online, whether movies, music, or restaurant ads, many companies (and academics) have found it worthwhile to develop so called recommender systems. A recommender system is a subclass of information filtering systems that seeks to predict the "rating" or "preference" that a user would give to an item (Wikipedia). Recommender systems have become increasingly popular in recent years and are extensively used, for example, in choosing what movies to watch, what music to listen to, what news to watch, which books to read, and which restaurants to visit.

The underlying logic in recommender systems can be categorized into collaborative-filtering approaches and content-based–filtering approaches (Waila et al., 2016). Collaborative-filtering approaches are based on the idea of building a model from a user's past behavior as well as other users' behavior (items previously purchased). The logic of incorporating other person's likes is that if other people found this item (or similar items) popular, so would you! Content-based–filtering approaches develop a set of characteristics that an item possesses (which you liked) to recommend additional items with similar properties.

Consumers generally appreciate recommender systems. However, we hesitate recommending them to filter news items that one sees. If an individual is solely or largely dependent on reading news in social media, as opposed to traditional media, recommended (filtered) by a system, the set of news offered becomes narrow, representing a very narrow worldview. We think that in such cases, the recommender systems should periodically suggest different types of news, to broaden the person's horizon! (Of course, we are assuming that a broader horizon would be better than a narrower one.) But what such news would be, and how to do it, may not be trivial. It seems that Facebook CEO Mark Zuckerberg's ideas are different regarding the development of Facebook. In a recent interview by CNBC Business News and Finance, he says that Facebook will change its algorithm so that users will see less public content from businesses or publishers and more posts from their friends.

The logic underlying recommender systems should be understandable to MCDM scholars, although such systems have traditionally been developed and studied by computer scientists and AI scholars. We urge MCDM scholars to develop better recommender systems. Both MCDM and recommender systems are about modeling user's preferences (Lakiotaki et al., 2011).

Voting advice Applications (VAAs) are online systems to help voters find worthy candidates to vote for in national, presidential, and regional elections. Such VAAs are highly popular in many European countries, where sometimes more than half of the electorate use them. They are based on both the candidates and the voters answering a set of questions concerning political preferences. The system (the algorithm) then finds the candidates and party, which are "closest" to the voter's political preferences. The development of such VAAs involves solving many MCDM/behavioral decision-making problems. The questions must be discriminating, and there cannot be too many of them. They must have proper Likert-scales to make distance measurement meaningful. What distance measure should one use? Are the questions of equal importance to voters or should importance weights be used? If yes, how are they determined? Are voters interested in voting for candidates who have a higher likelihood of becoming elected?

Jyrki Wallenius (2017) gave a keynote on this topic at the Ottawa MCDM Conference. They also have a paper detailing the development of their VAAs and its implementation in Finland (Pajala et al., 2018). We urge other scholars to further work on their respective country's popular VAAs. It is an important problem, and in particular, in multi-party, multi-candidate elections, voters benefit from the use of such support provided by VAAs by making them much more aware of what the candidates stand for.

1.3 BIG DATA (AND ARTIFICIAL INTELLIGENCE)

According to a recent issue by *The Economist*, companies' most valuable resource is data. Data is being continuously generated from various sources, including cash registers, mobile phones, and Internet sites visited by millions of people daily. There is a realization by the corporate world that they should better use this data to their (strategic) advantage.

Typical advertising and marketing agencies or departments do not know how to analyze big data, even though they realize its importance or potential. The need for people possessing analytics skills is high.

What role does big data play in advertising? In a nutshell, big data can be used to help create targeted and personalized campaigns that increase the efficiency of advertising or marketing. How is this done? Simply by gathering information and learning about user behavior. Many reward and loyalty programs are based on the use of consumer data. Recommender systems use past purchases or searches to make new recommendations. An interesting phenomenon is the use of social media by ad agencies. It is easy to document and share experiences as customer or consumer in social media. It is not uncommon that thousands of people read these posted reviews and are influenced by them. The world of social media offers interesting research opportunities to help businesses but also to understand human social behavior (Ghosh et al., 2017).

Another area where big data will find its uses is medicine or health care. Various monitoring instruments continuously generate data, as do human genome studies. They eventually lead to better preventive and actual care and more accurate diagnostics. An interesting problem from the perspective of MCDM is how to better incorporate patients' views on their own healthcare plans and treatment decisions. A more general level concern in health care is to make the system more efficient and more personalized. Healthcare decisions naturally have to deal with multiple criteria, and complex tradeoffs between cost, the quality of care, and even potential loss of lives. Wojtek Michalowski's (University of Ottawa) work is a good example of the type of impactful work a person with an Operations Research/MCDM background can do in health care. Jack Kitts (2017), President and CEO of Ottawa Hospital, gave a keynote at the Ottawa MCDM Conference, in part, based on Michalowski's collaboration with the hospital.

AI is a tremendously important field today. Part of the work uses Kohonen's neural nets (Kohonen, 1988). The idea is to build learning "robots," which could eventually make decisions on behalf of humans. An example is self-driving automobiles. Such "robots" need to be programmed to follow certain rules. They must make complex moral choices as well. Work is also currently being conducted to incorporate emotions into "robots." We ask, whose emotions? Our personal view is that we would hesitate to delegate decision-making powers in important matters to "robots," no matter how "intelligent" they are. We feel that humans should be in control of their own lives. AI is a good tool, but a dangerous master—something the ancient people said of fire.

1.4 THE SHARING (OR PLATFORM) ECONOMY

According to Wikipedia, *sharing economy* is an umbrella term with a range of meanings and is often used to describe economic activity involving online transactions. It grew out of the open-source community and referred to peer-to-peer–based sharing of resources and access to goods and services. The term is often used in a broader sense to describe sales transactions conducted via online market places (platforms). Online auctions are an example of such a market place, which have been around since late 1990s. Newer examples include the San Francisco-based taxi company, Uber, and an online market for housing, Airbnb. The clever innovation of Uber is that all that is needed is a platform where owners of cars and people in need of rides or deliveries can communicate. Uber is now operating globally in some 600 cities, without owning any vehicles. Airbnb is an American company which hosts an online marketplace and hospitality service for people to lease or rent short-term lodging, including vacation rentals, apartment rentals, homestays, or hotel rooms (Wikipedia). They currently have some three million listings. In the case of Airbnb what is needed is a platform where supply and demand for short-term housing meet. Another example of a sharing economy is crowdfunding and other peer-to-peer–lending sites, where private people (instead of banks) can lend money to people in need of money. Obviously, the interest rates are relatively high.

Our personal involvement with the sharing economy goes back to late 1990s, when we worked on developing a multi-attribute auction site, called *NegotiAuction* (Teich et al., 2001). We realized that price-only auctions were too simplistic and that auctions (transactions in general) need to include other aspects as well, such as quality and terms of delivery. Our *NegotiAuction* system was based on "pricing out" all other attributes besides cost. Today there exist many such commercial multi-attribute auction sites (Pham et al., 2015). More recently, we have investigated the success factors underlying crowdfunding campaigns (Lukkarinen et al., 2017). Generally speaking, many MCDM scholars are equipped with the skills to develop online platforms. We urge them to do so! There is a growing market for them. In sharing-economy platforms, some type of matching based on preferences is sought, where supply meets demand. The matching problem is a classic problem in economics (Pissarides, 2000). Lessons could be learned from economics as well as from MCDM.

1.5 CLIMATE CHANGE, CONCERN FOR ENVIRONMENT

Human-induced climate change is highly probable. B. Feltmate's (2017) keynote address at the Ottawa MCDM Conference dealt with it. The concern for the environment is almost universal. Most countries have signed the Paris Accord. Sustainable development is the keyword. When making decisions, corporations are increasingly forced to consider the impact of their decisions on the environment. If they fail to do so, consumers may boycott their products.

Generally speaking, environmental applications are probably the most common applications among MCDM studies. It naturally requires decision makers to consider multiple criteria and complex tradeoffs between them. See, for example, the book by Hobbs and Meier (2003). Another case in point is flood-risk management, an area, which is growing in importance because of climate change (deBrito and Evers, 2016). We believe that many models being used by various environmental authorities in the world may not be up to date in terms of the MCDM community's standards. We should increasingly get involved in helping model and solve problems related to the environment. It is our core business!

1.6 HOW IS MCDM CHANGING?

We have already seen the trend from multiple-objective optimization toward decision support. We are no longer so fixated on trying to find "optimal" solutions to problems, but supporting decision makers in many reasonable ways. The role of transitivity is probably eroding, as predicted by Fishburn (1991), although orthodox decision analysts do not see it that way. Heuristics are becoming more and more important. One good example is Evolutionary Multi-Objective Optimization (EMO), which consists of heuristic tools mimicking the survival-of-the-fittest ideas in nature (Deb, 2001). Although it is a relatively new field, it is doing great. Originally developed mainly for bi-objective problems, with the purpose of generating all approximately Pareto-optimal solutions, much recent research has focused on developing hybrid interactive-EMO approaches for multiple-objective problems.

The importance of the psychology of decision making, or behavioral decision theory, is being rediscovered. Three Nobel Prizes in Economics have been awarded to decision psychologists: the first to Herbert Simon[2]

[2] Obviously Herbert Simon is much more than one of the father's of behavioral decision theory. He is also regarded as the father of AI.

in 1978, the second to Daniel Kahneman in 2002, and the most recent to Richard Thaler (2017), whose work builds on Daniel Kahneman and Amos Tversky. We take a pragmatic view to the importance of behavioral issues in decision making. We think that the more realistic our tools are from a behavioral perspective, the better our chances to support individual decision makers. Hence, there is a need for improving the incorporation of decision psychologists' findings into our decision-support tools. Kahneman and Tversky's research takes us a long way. We also think that there is an increased awareness of the fact that situations vary and the needs of decision makers vary. In some cases there is a need for more formal analysis than in other cases. Sometimes, quick-and-dirty intuition may be all that is needed.

The Internet is changing the concept of who a "decision maker" is and what type of support he or she needs. We have largely been in the business of supporting corporate leaders and managers. How many corporate leaders are there in the world? A few million? But there are 4–5 billion consumers who shop online. Many of them could use some support when making purchasing decisions on the Internet. Such decision support must be targeted at masses; hence it must be simple. We think, in addition to complicated algorithms and decision-support tools, there is a need for developing simple tools to be used by the masses.

REFERENCES

Brechbuhl, H. World Economic Forum. https://www.weforum.org/agenda/2015/09/6-technology-mega-trends-shaping-the-future-of-society/.

deBrito, M. and Evers, M. (2016), "Multi-Criteria Decision Making for Flood Risk Management: A Survey of the Current State of the Art", *Natural Hazards Earth System Sciences*, open access.

Deb, K. (2001), *Multi-Objective Optimization Using Evolutionary Algorithms*, Wiley, Chichester, UK.

Feltmate, B. (2017), "Un-Natural Alliances: Financial and Ecological Expertise Must Align to Address the Contagion of Climate Change", *A Keynote at the 24th International MCMD Conference*, Ottawa, Canada.

Fishburn, P. (1991), "Decision Theory: The Next 100 Years", *The Economic Journal* 101 (404), 27–32.

Ghosh, A., Monsivais, D., Bhattacharya, K., and Kaski, K. (2017), "Social Physics: Understanding Human Sociality in Communication Networks", in *Econophysics and Sociophysics: Recent Progress and Future Directions*, Springer, Cham, Switzerland, 187–200.

Hobbs, B. and Meier, P. (2003), *Energy Decisions and the Environment: A Guide to the Use of Multi-Criteria Methods*, Kluwer, Boston, MA.

Kitts, J. (2017), "Is It Possible to Create a Sustainable Healthcare System in Canada?" *A Keynote at the 24th International MCMD Conference*, Ottawa, Canada.

Kohonen, T. (1988), "An Introduction to Neural Computing", *Neural Networks* 1 (1), 3–16.

Lakiotaki, K., Matsatsinis, N., and Tsoukias, A. (2011), "Multicriteria User Modeling in Recommender Systems", *IEEE Intelligent Systems* 26 (2), 64–76.

Lukkarinen, A., Teich, J., Wallenius, H., and Wallenius, J. (2017), "Success Drivers of Online Equity Crowdfunding Campaigns", *Decision Support Systems* 87, 26–38.

Pajala, T., Korhonen, P., Malo, P., Sinha, A., Wallenius, J., and Dehnokhalaji, A. (2018), "Accounting for Political Opinions, Power, and Influence: A Voting Advice Application", *European Journal of Operational Research* 266 (2), 702–715.

Pham, L., Teich, J., Wallenius, H., and Wallenius, J. (2015), "Multi-attribute Online Reverse Auctions: Recent Research Trends", *European Journal of Operational Research* 242, 1–9.

Pissarides, C.A. (2000), *Equilibrium Unemployment Theory*, 2nd edition, The MIT Press, Cambridge, MA.

Teich, J., Wallenius, H., Wallenius, J., and Zaitsev, A. (2001), "Designing Electronic Auctions: An Internet-Based Hybrid Procedure Combining Aspects of Negotiations and Auctions", *Electronic Commerce Research* 1, 301–314.

Waila, P., Singh, V., and Singh, M. (2016), "A Scientometric Analysis of Research in Recommender Systems", *Journal of Scientometric Research* 5 (1), 71–84.

Wallenius, J. (2017), "A Voting Advice Model and Its Application to Parliamentary Elections in Finland", *A Keynote at the 24th International MCDM Conference*, Ottawa, Canada.

CHAPTER 2

MCDA/M in Telecommunication Networks

Challenges and Trends

João Clímaco and José Craveirinha

CONTENTS

2.1 MOTIVATION

Telecommunication networks and technologies, as well as the services they support, have been and continue in a process of extremely rapid evolution. These trends, fostered by an exponential increase in offered traffic and a substantial demand for better and more advanced services, constitute a process of the greatest importance, not only in terms of technological advances but also regarding their quite significant impacts on the economy, keeping in mind the large investments involved on society as a whole. The evolution of these networks gives rise to a great variety of complex and multifaceted problems, typically involving multiple dimensions, often of conflicting nature. This means that the interaction between the complex socioeconomic environment of nowadays societies and the extremely fast evolution of new telecommunication networks clearly justify the interest in using multi-criteria evaluation in decision-making and analysis processes concerned with multiple activities of network planning and design. These factors, and the past research experience of the authors in some of these areas, laid the motivation for the topic of the present work, where we will seek to analyze and discuss trends and challenges of MCDA/M in relation to telecommunication network evolutions. A state-of-art review on applications of MCDA/M in telecommunication network planning and design was presented in Clímaco et al. (2016), in which papers up to 2012 were analyzed.

This work is organized as follows. In the next section we will present an outline of the more relevant evolutions of telecommunication network technologies and services, emphasizing more recent and foreseeable developments. Also in this section, an outline of major multi-criteria approaches and methods, all relevant in the addressed application areas, will be put forward. In Section 2.3 we highlight more recent or relevant works using multi-criteria models published in the context of routing methods and planning and design of telecommunication networks, including strategical and policy issues.

Finally, a discussion of future trends in these areas will be outlined, seeking to emphasize possible challenges concerning applications of MCDA/M to new technological platforms and services, as well as correlated methodological issues and challenges.

2.2 TELECOMMUNICATION NETWORKS AND MULTI-CRITERIA ANALYSIS

2.2.1 Evolutions of Telecommunication Networks: A Brief Overview

For a better understanding of the implications of network technologies evolutions in the emergence of a significant number of new sets of network planning and design problems, many of which may involve multiple criteria, we present a short overview of major trends and factors concerning such evolutions.

In historical terms, we can say that the most relevant telecommunication network evolutions have been centered on two major modes of information transfer: circuit switching (typical of classical telephone networks) and packet switching (typical of Internet). A historical milestone was the development of the transmission control protocol/Internet protocol (TCP/IP) suite, which enabled the very rapid expansion of the Internet in the 1980s, strongly accelerated in the 1990s through the release by the European Laboratory CERN (European Organization for Nuclear Research) in 1993 of basic Web technologies. Concerning the classical telephone networks, they rapidly evolved from the 1980s into ISDNs (integrated services digital networks ISDNs) enabling the convergence of different types of services and to broadband ISDNs (B-ISDNs), as a response to the rapid expansion of the demand for new data services and to more bandwidth-"greedy" services. These latter networks were based on the asynchronous transfer mode (ATM) technology, which was rapidly abandoned after 2000, as a consequence of the emergence of cost-effective multiservice Internet-based technologies, supporting the implementation of connection oriented services and advanced quality of service (QoS) routing and network-management mechanisms. In this context, we could refer to integrated services (IntServ), and differentiated services (DiffServ) technologies.

In recent years multiprotocol label switching (MPLS) and generalized MPLS (GMPLS), based on optical networks, have emerged, as more advanced base technologies for use in IP networks. In MPLS (Awduche et al., 2001), label switched paths are established, enabling traffic flows to be carried while ensuring various QoS requirements. Note that a fundamental reason for the success of Internet-based technologies, as basic communication transfer platforms, is the fact that they enable a high percentage of the capabilities of an "ideal network," in terms of supported services, at a significantly low relative cost, as shown in Handley (2006). These evolutions were supported, at the level of the transport infrastructure (transmission networks) by the development, especially in the last decade, of more advanced optical networks capable of making the most of the large bandwidths associated with the extremely low wavelengths that may be carried by optical fibers. This trend led to the deployment of wavelength division multiplexing (WDM) and dense WDM (DWDM) networks, the latter using tens of wavelengths on each fiber, enabling extremely large information rate transfers and an enormous traffic carrying capability with flexibility, provided by the introduction of wavelength conversion in the optical switches.

As for the transmission technologies (or "carrier technologies") based on optical fibers, the existing synchronous optical network/synchronous digital hierarchy (SONET/SDH) systems gave rise to new carrier technologies, such as Carrier Ethernet (Reid et al., 2008) and MPLS-transport profile (TP) (Niven-Jenkins et al., 2009), as cost-effective and functionally advanced alternatives. A major recent trend in transport technology evolution for optical networks is optical transport network (OTN) originally designed (see ITU [2009]), as the base transport system for SDH and subsequently extended for supporting IP and Ethernet. It has a multiplexing hierarchy from 1.25 to 100 Gb/s, capable of coping with very large bandwidths and providing advanced capabilities in terms of operations, administration, maintenance, and provisioning at wavelength level. The interplay between these technologies in the different functional layers of telecommunication networks enables the use of various network architectures such as IP/MPLS over WDM or IP-over-OTN-over-WDM, the latter being expected to reduce the needed router capacities and power consumption, also enabling a more efficient utilization of bandwidth, as a result of the advanced capabilities of the OTN switches. Another area where there have been extremely rapid evolutions concerns wireless networks, driven by the exponential increase in the demand for mobile data services, namely Internet access, since early 2000, where the annual increase rates were 60%–80%—*apud* (El-Sayed & Jaffe, 2002). The rapid evolution of third-generation (3G) to fourth-generation (4G) networks provided mobile broadband access to different mobile devices in many countries in a comprehensive and reliable IP-based network. These technologies enabled a quick shift from traditional telephone services, predominant in earlier technologies, to data services. Note that 4G networks are interoperable with existing wireless standards, enable significant improvements in QoS performance, and provide an extensive range of services, including applications like high-definition (HD) broadcast, video calls, and mobile TV, as well as a multitude of applications for entertainment, business, social networking, education, etc. For instance, according to a Cisco report (Cisco, 2017) 4G traffic accounted for 69% of global mobile traffic in 2016, although 4G connections represented only 26% of mobile connections in 2016. According to Tran et al. (2017) it is expected that mobile data traffic will continue to double every year in the near future. Another major factor driving the developments in wireless networks is the expected increase in mobile video traffic; according to a Cisco report (Cisco, 2016) this traffic already represented 55% of the total video traffic, and there are estimates of an annual growth rate of more than 60% (Tang et al., 2017). For example, at present, it is estimated in CTIA (2017) that 7% of US consumers watch mobile videos daily. Other important aspect should be mentioned, concerning new types of service demand, namely the great impact on telecommunication networks of the unprecedented evolution of cloud computing. This trend generates significant amounts of traffic flows of new type, namely dynamic "anycast" flows (i.e., from one origin to one of many possible destinations) and poses new challenges to the design of transport networks, as analyzed in Contreras et al. (2012).

These factors pave the way and justify the necessity for the next step in wireless technology evolution, the fifth generation (5G) networks (Monserrat et al., 2014). This is expected to provide important quantitative and qualitative advancements regarding increased bandwidth access (enabling new or better QoS data streaming services, including broadband services) and transmission latency (enabling more stringent requirements for real-time services). This will answer to the technical challenges raised by the fact that mobile users are subject (unlike users in wired networks) to time varying, significantly heterogeneous transmission channel conditions and variable availability of network resources. Note that International Telecommunication Union-Radio Communication Sector (ITU-R) and the Next Generation Mobile Network Forum have proposed ambitious objectives for 5G networks, such as access bit rates up to 10 Gb/s. The implementation of 5G will also require the need for coordination among various domains of telecommunication networks, namely involving wired networks, radio access, distributed processing, and service-related functions.

The mentioned expected evolutions in wired and wireless networks are also fostered by the development of the Internet of Things (IoT) in which a plethora of devices are equipped with electronic systems, sensors, and software, enabling to exchange data through the Internet. According to Ali et al. (2015), it is forecast more than 28 billion machine-type devices will be connected to the Internet by 2021, surpassing the number of expected human-centric connections. The IoT involves multiple interrelated technologies, and its use includes a multitude of applications, from remote smart-home control, intelligent transportation systems, grid automation, and remote health care to industrial automation (see, e.g., Zanella et al. [2014]). The convergence of IoT and cloud-computing technologies is also a developing trend, keeping in mind the limited resources of IoT devices and the fact that cloud servers can be used for data processing and storage. According to some authors (Perera et al., 2014), the technological and industrial-economic impacts of IoT make that it may be considered one of the main forces behind a fourth industrial revolution. There is a significant number of issues, risks, and design challenges raised by IoT technologies, namely of technical, economic, social, cultural, privacy, and security-juridical nature, keeping in mind the massive impact that these technologies, involving both human-centered and machine-to-machine–centered interconnections, may have in a near future.

A new technological paradigm that is expected to have a decisive impact in overcoming important limitations in the working and management of current network structures is software defined networking (SDN). The basic concept behind SDN is the separation between the network control logic and the underlying devices that actually implement the forwarding of traffic flows, achieved by direct control of various types of hardware through common management interfaces (Gallis et al., 2013). The introduction of SDN has quickly moved from small-scale data center/campus networks to large-scale carrier networks and is also being developed for application to 5G wireless networks (Rostami et al., 2017). This may be viewed as a major development in a wider trend directed to the "softwarization" of key network functions, based on the separation of the control plane functionalities from the data transport plane. This has major implications in terms of service provisioning flexibility and technical efficiency of networks, thence enabling a reduction in investment and operational costs.

Finally, it should be stressed the increasing relevance of multidimensional QoS and quality of experience (QoE; i.e., the multiplicity of performance measures as perceived by the end users, e.g., service availability or communication latency in a given service) issues in relation to the technological platforms. The QoS/QoE objectives and requirements are defined and have to be analyzed in the context of multi-service Internet-based technological platforms and involve the assessment of multi-dimensional QoS parameters and of the associated network control and traffic-management mechanisms. These issues have important reflexes in the type and nature of many new problems of network planning and design, e.g., concerning routing methods, the choice of alternative network architectures, or the evaluation of alternative policies involving socioeconomic factors. These issues as well as the main functional features of the networks, increasingly involving heterogeneous, interoperable technical platforms, increase the complexity and reinforce the multi-dimensional nature of many planning and design problems and of the associated decision analysis problems. The multi-dimensional nature of the aspects to be evaluated and the often conflicting nature of the criteria that should be included in the decision models associated with various instances and problems of the planning and design processes make it interesting and potentially advantageous in many situations to develop MCDA/M approaches in this broad area.

2.2.2 Multi-criteria Models: A Brief Overview

It seems clear that decision processes associated with telecommunication networks take place in a more and more increasingly complex environment characterized by a very fast pace of technological evolution combined with significant improvements in offered services. This trend is interrelated with drastic transformation of market workings and societal expectations. These key aspects of telecommunication networks evolution often involve multiple and potentially conflicting criteria. This is undoubtedly an area where various decisions of socioeconomic nature have to be made, but where, at the same time, the technological issues are of critical importance as pointed out by Nurminen (2003): "The network engineering process starts with a set of requirements or planning goals. Typical requirements deal with issues like functionality, cost, reliability, maintainability, and expandability. Often there are case specific additional requirements such as location of the maintenance personnel, access to the sites, company policies, etc. In practice the requirements are often obscure." This author, who collaborated with Nokia in the development of network planning and design models, makes it clear the limitations of single-criterion approaches. Nevertheless, he put in evidence the difficulties in the tuning of the parameters involved in mathematical programming models. He also drew attention to the fact that this issue is more difficult to tackle in multiple-objective formulations, once the procedures of preference aggregation by the decision maker(s) (DM(s)) require, in general, the specification parameters, for example, the determination of some sort of "weights," objective function thresholds, etc. However, this difficulty does not imply less interest in the development of multi-criteria approaches in this area, although it must be seriously taken into account. In fact multi-criteria models address different concerns of the decision process in an explicit manner, enabling the decision maker(s) to grasp the possibly conflicting nature of the considered criteria, so that he or she may tackle the compromises that have to be made to obtain "satisfactory" solutions. Of course, these difficulties are extensive to multi-attribute models.

When different and conflicting criteria are at stake, the concept of optimal solution is replaced by the concept of nondominated (also known as Pareto optimal or efficient) solutions set. This includes only feasible solutions for which no improvement in any criterion is possible without worsening at least one of the remaining criteria. In general, we can say that multi-criteria choicees seek to obtain one or more nondominated (or at least approximately nondominated), solution(s) considered as satisfactory by the decision maker. Note that choosing the method that is used for aggregating the decision-maker preferences is also multi-criteria in nature. Beyond the difficulty previously mentioned concerning the specification of parameters in the developed models, we should access whether there is the possibility of using interactive procedures, especially taking into account the required speeds of calculation for a given application. This means that an interactive procedure cannot be used if the calculations, in an interaction, are too slow, with respect to the envisaged application. Furthermore, in various telecommunication network decision problems (e.g., in many routing methods), no more than a few seconds (sometimes much less) can be accepted for finding a final solution to be implemented. These are cases in which interactive procedures cannot be used in practice. All these factors, including cognitive and technical aspects, are at stake, so that, in many cases, the quality of the selected solutions may be compromised.

Of course it is important to address, in a wider perspective, which multi-criteria model is more adequate to each case. We referred mostly to mathematical programming models, that may be linear, non-linear, and additionally, may have, or not, a specific structure. In contrast, there are other types of models, here designated as multi-attribute decision models, that also have been subject to significant developments, also including applications in telecommunication issues. Although in multi-criteria mathematical programming models, it is assumed that the set of feasible solutions and alternatives is defined implicitly through the constraints, in multi-attribute models a small and discrete set of alternatives is specified explicitly. The alternatives in this set are then analyzed with respect to multiple criteria (or attributes). Note that in this type of models it is possible to carry out a more detailed evaluation of the alternatives, considering a bigger consistent family of criteria, and this can be done without implying a computational explosion. Nevertheless, in many situations of network design and planning (namely in typical routing and facility capacity calculations), this implies a reductive point of view, which may not be realistic because it does not enable a proper exploration of the decision space. As will be illustrated, in the highlights of some studies in this area, the complementary utilization of both types of approaches can be advantageous in some specific problems.

Multi-attribute models, in the so-called American School a multi-attribute utility function (based on multi-attribute utility theory), may be constructed either linearly or non-linearly (depending on the problem) (Keeney & Raiffa, 1993). In the case of the analytical hierarchy process (AHP), this can be viewed as a particular branch of the American School, that involves the identification of a hierarchy of interrelated decision levels (Saaty, 1980, 1994a, 1994b). An alternative methodology is the so-called French School, the basic principle of which is the introduction of partial orders; that is, outranking relations are involved. This means that no more complete comparability of alternatives and transitivity relations are obtained. In conclusion, these methods are less demanding than the former, concerning the fixation of parameters, but, on the other hand, in general, they do not allow a complete ranking of alternatives; hence, they do not guarantee the principle of optimality (i.e., neither transitivity nor full comparability are verified). Therefore, their results are less conclusive with respect to the aggregation of the preferences of the decision maker.

As the most relevant example of the French School approaches, we can mention the ELECTRE family of methods (Roy & Bouyssou, 1993, (Figueira et al., 2016). Depending on the problem, the purpose is the selection of the most preferred alternative, the classification or the ranking of alternatives.

More recently, mathematical programming and multi-attribute approaches basing the preference aggregation in inductive rules have been developed. Namely, the approaches rooted in an adaptation of the rough sets concepts must be emphasized (Slowinski et al., 2012).

Concerning the approaches dedicated to multi-criteria mathematical programming models, attention should be paid to the dimension of the real problems we are dealing with and, many times, as noted previously, there is the necessity of a rapid execution. We would like to note that, in many situations, the mathematical programming models to be used have a network structure, and in some of these cases, there are very efficient specific exact algorithms for solving even big instances. This is the case for models involving multi-criteria shortest path problems (see, e.g., Clímaco & Pascoal, 2012).

However, in most of the situations, this is not the case. Therefore, it is often necessary to use heuristics and meta-heuristics for resolving these models in acceptable computational times, namely when online (and specially real-time) calculation methods are to be implemented. In particular, the development and application of multi-objective evolutionary algorithms is remarkable and, as we will show in our summary of some papers, these methods have also been applied to some problems of telecommunication planning and design.

Furthermore, another key issue has to do with the treatment of the uncertainties in various instances of the models. In particular, in many models, the uncertainty associated with traffic flows offered to the network is of great importance. The representation of this uncertainty is a task with two major aspects: the use of adequate stochastic models (often mere approximations) for the traffic flows, in the context of the model, and the determination of estimates of the statistical parameters of the stochastic submodels. Uncertainties or imprecisions inherent to other quantities involved in the multi-criteria model, that may be of different natures, for instance data collection or modeling of preference aggregation (see Bouyssou [1990]), are also relevant issues in this regard.

Remember that multi-criteria approaches enable, in these conditions, the identification of the set of criteria associated with the stable part of the decision-makers' preferences, so that a further aggregation of their preferences is left for further analysis. So, in many cases, the output of the multi-criteria analysis is not a solution but a set of satisfactory solutions, in the context of the used model. Thence, an *a posteriori*, more detailed analysis of those solutions (namely, having in mind characteristics which were not initially included in the model) is advisable.

In the next section of this text, an outline of more relevant works using multi-criteria models, published in the context of planning and design of telecommunication networks, as well as in the context of socioeconomic implications of telecommunications evolution, including strategical issues, is presented.

2.3 HIGHLIGHTS OF APPLICATIONS OF MCDA/M

In this section we will present highlights on recent applications of MCDA/M telecommunication network planning and design problems including strategic planning and policy issues. For a better understanding and facilitation of the analysis of the problematic areas where there has been a cross-fertilization between MCDA/M and telecommunication networks we will consider three areas of decision support and optimization issues, each corresponding to a subsection. The first area is focused on highlights of recent routing models, an area where there has been a great increase in contributions using various types of multi-criteria–based models. The second area refers to network planning and design issues and papers that present multi-criteria modeling approaches dealing with socioeconomic evolutions associated with specific telecommunication network problems and the third area includes strategic planning and telecommunication policy evaluation problems. It should be strengthened that there is no sharp frontier between these areas, noting, for example, that network design includes implicitly or explicitly some routing submodel and that most models of network planning and design involve, either directly or implicitly, economic or social aspects.

2.3.1 Routing Models

In the general context of planning and design processes, routing is a key network functionality that may be considered as an integral part of the network operational planning decision process. It is strongly related to other planning activities, namely network structure design (that includes topological design and equipment capacity calculation) and traffic network management (a top-level network functionality aiming at a dynamic global optimization of traffic flows throughout the network, having in mind information on the currently available resources and offered traffic). Routing models are essentially concerned with the calculation and selection of a path or set of paths from an originating node to one (several) terminating node(s) (considering that the network representation is a connected graph the arcs of which have a limited transmission capacity), seeking to optimize certain objective(s) and satisfy certain technical and economic constraint(s). Routing solutions have a strong impact on network performance, namely in terms of traffic carried and resulting QoS levels and cost/revenues of the network operator(s).

An important class of routing methods, other than the most common *point to point (or unicast) routing*, involves the calculation of several paths simultaneously, between two nodes or between two sets of originating and terminating nodes. These methods correspond to a class of routing problems designated, in general, as *multipath routing* problems. A specific type in this class (which may be designated as point-to-point multipath routing) refers to routing models with reliability requirements or objectives, or resilient routing, in which an active path and a back-up path (which will be used in the event of failure in the active path), have to be computed for each pair of origin-destination nodes. Another type is *multicast routing* in which a set of paths has to be calculated from the originating node to a set of destination nodes—point-to-multipoint routing. This is the type of routing for the distributional services supplied by a certain service provider or interconnecting two subsets of network nodes, a multipoint-to-multipoint routing model, for example, in teleconferencing services in Internet. Assuming all the nodes have to be interconnected, the multicast routing problem is designated as *broadcast routing* and is usually formulated as a spanning tree problem. If the set of destination nodes is a proper subset of the set of network nodes, the corresponding multicast routing problem is typically formulated as a Steiner tree problem, where the destination nodes and the originating node are the terminal nodes.

Routing problems may have different natures and often a multiplicity of formulations, depending on fundamental aspects, namely: the mode of information transfer; the type of service(s) associated with the routed connection demands (e.g., a telephone call, a video-service, a data stream transfer, a wavelength assignment); the considered level of representation of the network (typically, at least two levels may be considered: the physical or transmission network and the logical or functional network) on which the routing problem is formulated; and the main features of the routing paradigm (e.g., whether it is static or time varying according to traffic fluctuations or network conditions in a given time scale). The network technical entities that actually implement, at a lower level of functional network representation, the routing function are the routing protocols, critically interrelated with the network technological features.

The rapid technological evolution in the late nineties, associated with the increase in the demand for new communication services, mainly Internet-based services, implied the necessity of developing multi-service networks capable of dealing multiple, heterogeneous QoS metrics. As noted previously, in these networks different classes of services are specified (in the context of given technological platforms), which have different requirements of QoS. The performance of these networks is naturally a function of the degree to which such requirements are achieved and is also expressed in terms of global network measures such as mean traffic carried, blocking probability or average delay. This led to a routing paradigm in telecommunication networks designated as QoS routing. This type of routing methods involves the calculation and selection of chain(s) of network resources along one or multiple feasible paths from origin to destination, satisfying given QoS requirements. These requirements are dependent on traffic features associated with service classes, so that the associated QoS routing algorithms need to consider distinct metrics (Lee et al., 1995). These routing models typically seek to optimize some metric(s) such as delay, cost, number of edges of a path or loss probability, and the other metrics are treated as constraints. In this context the path calculation problem is typically formulated as a shortest path problem with a single objective function, which is either a single metric or a function of different metrics, and QoS requirements are included as constraints, that is, it leads to a constrained optimal path problem.

This type of models (usually designated, in the telecommunication literature, as constrained QoS routing) may be considered, as proposed in Clímaco et al. (2016) as a *first tentative of MCDA/M modeling*. This is justified by the well-known principle that a possible approach in multicriteria model analysis is the transformation of the initially considered objective functions into constraints, excepting one objective function that is optimized. The solution obtained with this procedure is (in adequate conditions) necessarily a nondominated solution for the original multi-criteria model; moreover, it is possible to obtain different nondominated solutions by varying the value of the second member of the constraints (see Steuer [1986]). This posture, concerning the characterization of approaches, which are explicitly multi-criteria, was also adopted in Wierzbicki and Burakowski (2011). These authors proposed a conceptual framework for the development of explicitly multi-criteria modeling approaches in the context of QoS routing in IP networks

Therefore, in the present highlights of papers, we will refer only to recent contributions on models which are *more explicitly multi-criteria*. Moreover, we think there are significant advantages in approaching many routing problems in modern telecommunication networks, through explicitly multi-criteria formulations. This type of modeling approach to such problems is potentially advantageous, although we cannot ignore that in many instances of routing design, the solution to be implemented in a given technological context has to be calculated in a short time, that may range from a small fraction of a second (typically up to tens of ms) to a few seconds. In these cases, as noted, there is no possibility of using interactive resolution methods, thence leading to the necessity of developing automated path calculation and selection procedures. Nevertheless, there are many situations in which this limitation does not apply, namely in static routing methods, in transport networks where transmission paths are maintained for relatively large time periods or in various types of dynamic routing methods, where the input parameters of the routing algorithm are estimated in advance (e.g., considering node-to-node traffic intensities or current link bandwidth occupations, in different time periods), cases in which an interactive procedure could be used to select the routes (for every node pair) to be memorized in routing tables assigned to every router, to be up-dated only after many minutes, when new transmission route(s) have to be calculated. This means that there are many routing models, considering multiple criteria, some of which will be illustrated next, where it is possible the conciliation of automatic path calculation procedures with some flexibility in the form of preference aggregation. For this reason and the possibility of using interactive procedures, in various routing problems (not involving real-time/short-time routing decisions), adequate multi-criteria approaches enable the grasping of the compromises among different and conflicting criteria, also taking into account various QoS requirements. Moreover, such approaches enable a consistent comparison among distinct routing possibilities, in the context of a certain routing principle.

In our highlights of recent contributions, illustrative of the application of MCDA/M, we will consider "clusters" of routing problems of different types (in italic) and, for each type, a classification acronym for each reference, dedicated to the papers outlined in this work, according to the MCDA/M approach or method used in the modeling or resolution method.

Having this in mind, we consider a tentative classification of the used multi-criteria models and resolution methods, according to the following types and corresponding acronyms: (i) simple weight additive models, (SWAM–namely models where there is an *a priori,* direct or indirect, specification by the decision makers of weights assigned to the each criteria; (ii) multi-criteria network flow programming (MNFP); (iii) multi-criteria shortest path models (solved with exact algorithms) (MCSP); (iv) multi-objective integer linear programming–based formulations (MILP); (v) multi-objective nonlinear programming based formulations (MONLP); (vi) multi-criteria minimal spanning tree models (solved by exact algorithms) (MMST); (vii) multi-criteria heuristics (MH); (viii) multi-criteria metaheuristics (MMH); (ix) outranking methods: (ELECTRE) and (PROMETHEE) methods; (x) Analytic Hierarchy Process and extensions (AHP-E); (xi) multi-attribute utility theory based methods (MAUT). Note that under the classification MH we may find quite different techniques, from heuristics based on simple empirical enumeration rules of generation of feasible solutions, with elimination of the dominated ones, to dedicated heuristics, seeking to explore properties of the problem, often based on exact subalgorithms for generating candidate solutions. Also under the classification MMH we may find quite different procedures, from simulated annealing to evolutionary algorithms of various types.

We will begin by considering some *multi-criteria routing models for Internet.*

The paper by Girão-Silva et al. (2012) (MH) describes a dedicated heuristic, using a Pareto archive, for solving a complex hierarchical multi-objective routing model in MPLS networks with two service classes, formulated as a multi-objective network-wide optimization model (characterized by the fact that the objective functions of the route optimization problem depend explicitly on all traffic flows in the network), with stochastic objective functions, including fairness objectives; the developed heuristic is ultimately based on a bi-criteria shortest path subalgorithm, using as path metrics implied costs and blocking probabilities.

The paper by Girão-Silva et al. (2015) (MNFP) presents a multi-objective routing model for MPLS networks, considering multiple service types and traffic splitting, using a network-flow approach; the routing problem is formulated as a multi-objective mixed-integer program where the considered objective functions are the bandwidth routing cost and the load cost in the network links, with a constraint on the maximal splitting of the service bandwidth demand; two different exact methods are developed for obtaining nondominated solutions, one based on the classical constraint method and another based on a modified constraint method (Messac et al., 2003).

In Girão-Silva et al. (2017) (MILP) the authors propose a multi-objective resilient routing model for MPLS networks with multiple services and path protection, where the considered objectives are route cost and load cost; the routing problem is formulated as a bi-objective integer program, in the context of a network-wide optimization approach using a link-path formulation; an exact method based on the classical constraint method for solving multi-objective problems is used for obtaining all nondominated solutions, given the set of feasible node disjoint path pairs.

The paper by Bhat and Rouskas (2016) (MH) describes a new type of *routing model assuming marketplaces* of dynamically supplied "path services" that considers as objectives to be optimized, cost and expected delay, and includes various QoS requirements; users are supposed to choose from a set of "path services" offered by multiple competing network providers, which are feasible for given time intervals; the authors propose a dynamic programming heuristic for solving the associated multi-criteria constrained shortest path formulation for certain time windows.

Aissanou and Petrowski (2013) (MAUT) propose a MCDA/M model for route selection by an autonomous system in a dynamic data routing network, considering as criteria to be optimized, packet delay, and loss rate; the model uses a set of nested "quality boxes" in the criteria space, for defining an utility function; a learning heuristic procedure is proposed to configure the boxes based on subjective quality assessments provided by users, considering an application to *wireless ad-hoc networks*.

Thaalbi et al. (2013) (based on AHP-E) propose a multi-attribute model for route selection in a multipath dynamic routing process in mobile ad-hoc networks, the criteria being delay, jitter, packet loss rate and data rate, considering multiple service classes; the multi-attribute decision procedure used for selecting "best quality routes" was proposed in Savitha and Chandrasekar (2011) and is based on AHP.

Wireless sensor networks (WSNs) are composed of sensor nodes that are installed with the objective of gathering real-time information of certain type in a given area, so that the associated data are forwarded to a special node, the sink node , which is an area where multi-criteria routing models have been recently proposed. Most proposals aim at the introduction of "fast" heuristic procedures of path calculation in the routing protocols, taking into account several criteria to be optimized. Sahli et al. (2012) (SWAM) describe a generic routing framework concerning the criteria to be addressed and discuss a form of additive aggregation based on technical features of the used routing protocol.

Bhunia et al. (2014) (SWAM) present a multi-criteria routing model for WSNs, considering as objectives residual energy of a node, frequency count of packet transmission via a node, value of frequency count of packet transmission via a node, number of hops counted from the sink node; a heuristic routing procedure based on additive aggregation of criteria using various weight sets empirically evaluated in terms of the resulting packet loss ratio is proposed. The same type of modeling approach is presented in Das et al. (2015) (SWAM) but considering a heuristic based on a weight product calculation (with weights assigned to each criterion) where the weights are chosen by a "weight rating" method.

Also Suh et al. (2015) (SWAM) describe a multi-criteria routing procedure for WSNs, considering distance, queue length, and residual energy of each node; the model uses a concept of "virtual potential field-based energy" routing and a weighted normalized decision matrix for choosing the next node to be selected in the path.

In a model, shown in Rehena et al. (2017) (MH), the two criteria, in WSN routing with partitioned sink nodes, are the distance of the node from the sink and the remaining energy of a node; a heuristic procedure is used, based on the calculation a decision matrix the elements of which are obtained from an utility function, involving those criteria, the value of which is associated with the choice of the next node to be selected in the path. Note that almost all these routing models for WSNs use a "step-by-step" path calculation heuristic, where the next node in the path is chosen through a multi-attribute model based on the construction of a performance matrix at each step of the procedure.

In Bueno and Oliveira (2014) (MMH) a multicast routing model is formulated as a multi-objective Steiner tree optimization problem; the objective functions are the tree cost, mean end-to-end delay to the destination nodes, maximum end-to-end delay, number of arcs, and maximum link utilization; a metaheuristic based on the Strength Pareto Evolutionary Algorithm (Zitzler et al., 2002) is used as resolution procedure, and three variants of the heuristic are tested.

A bi-criteria *resilient routing model* (MCSP) for transport networks is described in Gomes et al. (2012) seeking the calculating of a bi-criteria active path (in terms of minimal load cost and hop count) with a maximally disjoint protection path; an exact resolution method is described, which is based on a k-shortest path algorithm, applied to the convex combination of the two objective functions, hence enabling to obtain all supported and unsupported nondominated solutions. Gouveia et al. (2016) (MILP) present a multipath problem in the context a general resilient point-to-point routing model by considering a lexicographic optimization formulation; the aim of the formulation is to minimize the number of service (or active) paths with the worst number of hops, such that each connection demand is routed through a set of node disjoint service and backup paths, all with a bound on the number of arcs; and integer linear programming formulations are specified and tested for obtaining exact solutions.

The paper by Gomes et al. (2016) (MH) proposes a lexicographic approach to the point-to-point resilient routing problem in GMPLS networks; the model involves the calculation of pairs of paths seeking to minimize, lexicographically, the number of common nodes, the number of common arcs, the number of common shared risk link groups (SRLGs; i.e., sets of arcs that share a common risk) and the path pair cost; two heuristics for solving the problem are developed and its performance evaluated with reference test networks.

The reference Craveirinha et al. (2013) (MMST) describes a bi-criteria minimal spanning tree routing model for broadcasting messages or defining *overlay networks* over a MPLS network; the considered objective functions are the total load balancing cost and an average upper delay bound on the arcs of the spanning tree. An exact algorithm is used for the calculation of all supported nondominated solutions and one of such solutions is selected by a method based on the approach in Gomes da Silva and Clímaco (2007). In Craveirinha et al. (2016) (MMST) a bi-criteria optimization model for constructing resilient overlay or broadcast networks based on spanning trees over WDM optical networks is presented; the objective functions are the minimization of the total number of different SRLGs of the tree (hence, seeking to maximize reliability) and the minimization of the total bandwidth usage cost; the formulated problem is solved by an exact algorithm that is an extension of the minimal cost/minimal label algorithm in Clímaco et al. (2010), enabling the whole set of nondominated solutions to be calculated; and methods for selecting a final tree structure, in various practical decision environments, are put forward.

In Esteves and Craveirinha (2013) (MONLP) a stochastic bi-criteria problem, for calculation of the *allocation of servers* in a multi-dimensional Erlang loss system, considering a max-min criterion of equity in the blocking probabilities and the maximization of the total traffic carried by the system, is formulated. An exact algorithm for traveling on the Pareto frontier in the objective function space, based on a Newton-Raphson method, is also described.

2.3.2 Network Planning and Design

Network planning and design designates a vast area of activities, dealing with short- and medium-term network problems that are focused on the location, interconnection layout and dimensioning of transmission systems (cables, optical fibers, radio, and satellite links), and other facilities such as switching units, traffic concentrators, routers, or mobile stations. Operational planning usually refers to short-term network design, often encompassing network management, maintenance, and related activities. As for strategic planning, it deals with the development, analysis, and evaluation of scenarios of qualitative and quantitative network expansion, focused on medium- to long-term periods, taking into account the traffic growth, the demand for new services, the introduction of new technologies, and economic objectives. It must be noted that the frontiers between medium- and long-term planning are often blurred. At the highest level, strategic planning also concerns, explicitly, telecommunication government policy and socioeconomic issues. Note that this type of strategic decision problem involves a multiplicity of factors, some of which cannot be directly represented by an economic indicator.

Many network planning and design models seek to express different aspects of the associated optimization problems, involving, in reality, multiple requirements and often conflicting objectives, in terms of economic measures to encompass those aspects in a unique objective function. These models lack, in most cases, to capture explicitly the various and possibly conflicting aspects arising in evaluating network design solutions and network expansion policies. That is why MCDA/M models, by enabling an explicit consideration of technological, economic, and social aspects, allow the decision makers to tackle the conflicting nature of the objectives and analyze the trade-offs that have to be made, keeping in mind to choose a satisfactory solution.

Only in specific problems of network planning and design there have been recent proposals of multi-criteria modeling. Here we present highlights of some recent and significant papers in these areas. The same system of classification of papers, according to the MCDA/M methods used in the resolution approach and described in the previous subsection, is used.

The paper by Bezruk et al. (2012) (SWAM) describes a generic multi-criteria system-optimization approach for network design, seeking to obtain Pareto optimal variants of the network design solution; the approach is applied to the design of a *cellular wireless network* and the resolution method, after generating a set of permissible variants of the system, obtains nondominated solutions based on the optimization of a convex combination of the objective functions, the form of which is determined with the use of some additional information obtained from a decision maker.

Concerning the design of wireless networks Statnikov et al. (2013) (MH) describe a multi-criteria optimization model of cellular networks with seven quality of transmission-related parameters and two variables per cell (transmitter power and electrical tilt); the authors use the "Parameter Space Investigation (PSI) method" in Statnikov and Statnikov (2011) for obtaining nondominated solutions through an interactive search.

A problem of design of transmission systems and the *optimization of bandwidth allocation* is described in González et al. (2016) (MMH) by formulating a multi-objective optimization model for fractional frequency reuse in mobile wireless networks, considering as criteria to be optimized: system average bandwidth capacity, cell edge bandwidth performance, and energy consumption; an evolutionary metaheuristic is used for obtaining an approximation to the Pareto front of the formulated problem.

The paper by Shi et al. (2014) (AHP-E) describes a multi-attribute model for application of *countermeasures against malicious attacks* to nodes of mobile ad hoc networks (MANETs), for example, in military, emergency, or mining operations using a cluster-based strategy; the AHP methodology is used to choose "cluster head" nodes, which are supposed to implement the countermeasures by weighting the three selected technical criteria.

A multi-attribute model for *global performance evaluation* of IP-based networks (under different traffic loads or for comparison of networks with the same traffic) is proposed in Chen et al. (2014) (ELECTRE); from a network performance matrix, with tens of QoS parameters measured in different time periods, a maximizing deviation method based on ELECTRE principles (Chen & Hung, 2009) is used to determine the attribute weights; this leads directly to a ranking of network alternatives based on the resulting values of an additive value function.

The social penetration of communication technologies and services and the consequent socioeconomic implications justify why they are now in the agenda of various areas of science, philosophy, and politics. In fact, their present relevance is remarkable, and the future trends have still, in many aspects, unexplored dimensions.

It is clear that the associated analysis and decision problems are multidimensional, and it seems that multi-criteria models can be helpful tools for decision aiding in this domain. However, as these issues are relatively new, evolving very fast, and requiring also rapid options, the number of studies involving multi-criteria tools is still quite limited.

Next we make an outline of some relevant or more recent works dealing with these issues, while drawing attention to the used multi-criteria approaches. In the section dedicated to future trends, we will try to foresee auspicious future trends concerning applications of MCDA/M.

The use of multi-attribute models in telecommunications planning, as far as we know, has been mainly proposed for application in models studying interactions between telecommunication evolution and socioeconomic issues, as analyzed next. As we shall see, although different multi-attribute methods have been used, in most cases, AHP was the chosen method. Moreover, in some cases, mathematical programming approaches have also been proposed.

The references to the studies, reported hereafter, were done taking into account the type of problems in network planning and design, involving socioeconomic aspects, addressed in the papers and the type of multicriteria analysis method used by the authors. Note that the strategical issues are dealt with in subsection 2.3.3.

The paper by Mohanty and Dabade (2015) (MAUT) presents a real case study focusing on *supplier selection* to an Indian telecom service provider using an AHP technique.

In Wojewnik and Szapiro (2010) (MH) a model for *pricing of telecommunication services* is proposed; a heuristic procedure for interactive multi-criteria optimization involving fuzzy coefficients is presented.

The authors in Uygun et al. (2015) (AHP-E) describe a MCDA/M model for evaluation and selection of an outsourcing provider for a telecommunication company using a fuzzy multi-criteria approach (analytic network process [ANP]). Note that ANP is a generalization of the AHP methodology, in which hierarchies are substituted by networks that enable the modeling of feedback loops.

In Abourezq and Idrissi (2015) (ELECTRE) a multi-criteria model for searching and *selecting cloud computing services*, including criteria to be optimized, such as price, the bandwidth, and multiple QoS constraints, is presented; an outranking method, ELECTRE IS (Figueira et al., 2016), is used for solution selection.

In Adebiyi et al. (2015) (AHP-E) the authors describe a model of analysis of the *behavior of subscribers* concerning retention to a given operator and apply it in the Nigerian mobile telecommunication networks; AHP is applied.

Pereira and Bianchini (2013) (AHP-E) present a multi-attribute model for analyzing the major factors that determine the dissatisfaction of clients of mobile network operators in Brazil; a AHP method is developed for ranking of those factors, keeping in mind to reduce the number of complaints.

In Bentes et al. (2012) (AHP-E), a multi-attribute model for organizational performance evaluation of a Brazilian telecom company is presented; the MCDA/M model combines the balanced scorecard (BSC) method in Kaplan and Norton (1996) with AHP for the ranking of performances of functional units of the company.

2.3.3 Strategic Planning and Policy Issues

In this subsection we refer to studies that may be considered as focusing on strategic planning and telecommunication policy issues.

The study in Keeney (2001) (MAUT) addresses the issues concerning the construction of a value model dedicated to decision processes in *telecommunication company management*; the author pays particular attention to the structuring of objectives, taking into account both qualitative and quantitative aspects and considering the use of multi-attribute utility functions.

Colson et al. (2006) (PROMETHÉE) compare the performance in a determined period of telecom operators in four Maghreb countries. They propose the use of a data envelopment analysis (DEA) tool and a well know MCDA method (i.e., the PROMETHÉE II) to rank the countries. The study is done for three subperiods between 1992 and 2001. The authors consider the service technical-economic performance and the operators performance.

Grzegorek and Wierzbicki (2012) present an interesting use of multicriteria evaluation/ranking tools in the study of the social penetration of information society technologies in the framework of supporting regional policy. Of course, the scope of the study includes the communication technologies penetration, but it also has a broader scope. The authors make an overview of the available indexes and, emphasizing that an aggregation is always necessary to obtain a ranking, say that the use of classical additive aggregation is subjective. That is the reason why they propose a so called "objective ranking" procedure. Instead of eliciting weights from the decision actors, they just use statistical parameters to enable the aggregation. Note, as it is admitted by the authors, the method is not fully objective because it depends on the options made for those calculations.

The paper by Mfupe et al. (2017) (PROMETHÉE) presents a modeling approach for formulating a regulatory framework to govern the spectrum utilization by wireless networks that are based on the dynamic spectrum access (DSA) technique; for the evaluation of the DSA management policies of regulatory authorities, a multi-attribute analysis model with 11 criteria, including socioeconomic objectives, is presented and tackled with a PROMETHÉE method in Brans and Mareschal (2005).

Finally, we consider two papers not applying explicitly MCDA/M methods but based on some multi-criteria concepts.

The first one, by Desruelle and Stancik (2014), makes a descriptive comparison of the six principal world players in manufacturing and in services creation, concerning information and communication technologies; the authors consider value added, business expenditures in R&D (BERD), BERD intensity, and labor productivity.

The second one (Gerpott & Ahmadi, 2015) is an interesting paper regarding indices for accessing nation "telecommunications development"; the work justifies why and how they propose a new composite metric, aggregating 11 diversified indicators.

2.4 FUTURE TRENDS

Next, we will present an outline of foreseeable research trends and topics in some areas of network planning and design, including topics where it seems likely that more opportunities and challenges for MCDA/M may arise. For facilitating the presentation and help in systematizing the research topics, these trends will be organized in three parts: routing models, network planning and design and models studying interactions between telecommunication evolution and socioeconomic issues, and finally, strategic planning and policy issues.

2.4.1 Routing Models

We will discuss future trends concerning routing models, around topic clusters, separated by hyphens, with the application feature, common to topics in each cluster, indicated in italic.

- As a first topic we would like to note that operations research (OR)-based models in network planning and design usually consider a network representation through a capacitated graph and a matrix of node to node offered demand, or if we wish to have a complete representation of traffic flows (of a stochastic nature) and of routing methods, a more general representation, through a "teletraffic network," composed of several mathematical and other logical entities (see e.g., Craveirinha et al. [2008] and Clímaco et al. [2016]), should be used. Nevertheless, the nature of real telecommunication networks is even more complex because they are organized, from a functional, operational, and management/control points of view in several interrelated layers, leading to the necessity of considering them as *multilayer networks.* An Internet network, for example, even in a limited national area, has at least three layers, namely the physical infrastructure (or physical layer, including coaxial cables, optical fiber cables, and microwave links), the router topology layer (corresponding to the logical layer), and above this, the third layer, where application level and social network flows can be represented (see, e.g., Sprint service provider in the United States [KU, 2012]). This problem also raises difficult modeling issues as far as routing and network design models are concerned, keeping in mind the very great complexity of these structures and the interrelations between the various layers. A useful tool in this respect is multilevel graphs (Çetinkaya & Sterbenz, 2013), a mathematical representation of these networks, consisting of various graphs, one for each layer, corresponding to a level of the graph, such that the set of all nodes in a higher level is a subset of the set of all nodes in the immediately lower level and the nodes that are not connected in a lower level are equally not connected in the higher level. These aspects should be taken into account in OR-based models in general and in multi-criteria models in particular. Furthermore, we think, as noted in Rak et al. (2015), that multi-criteria routing approaches are potentially advantageous in the context of multilayer networks and pose interesting challenges. In fact, beyond the intrinsic advantages of multi-criteria routing approaches, already discussed in Section 2.3.1, and concerning resilient routing with protection, the development of such multi-criteria routing models could enable a consistent treatment of the trade-offs between various metrics associated with different routing and protection options in each layer, in different failures scenarios. This would require tackling the difficult issue of decomposition of the routing optimization model. In fact, the nondominated solutions of the routing optimization problem formulated for the physical transport network (lower level, e.g., a OTN optical network) would correspond to pairs of light-paths, such that each light-path may correspond to different possible paths, in the following layer (e.g., MPLS-TP). The routing optimization model would also be multi-criteria in this layer, so that a complex problem decomposition is at stake. Note that a first approach to a multi-criteria routing optimization framework for MPLS networks with a hierarchical structure based on a three level hierarchy of objective functions focused on global network objectives, service objectives, and micro-flow QoS objectives was earlier proposed in Craveirinha et al. (2008). However, a second kind of hierarchical modeling approach, suitable for multilayer networks, would involve a hierarchy of the routing optimization formulation, that is, concerning the application of classical optimization methods (see, e.g., Findeisen et al. [1980]) to multilevel routing as suggested in Wierzbicki and Burakowski (2011). This involves, in single-criterion approaches, the decomposition of routing problems into routing subproblems, concerning different domains, and then the composition of solutions of these subproblems to seek global optimality. The adaptation (in the first case, noting that it is already a hierarchical multi-criteria approach) or extension, in terms of multi-criteria optimization (in the second case) of these two types of modeling approaches, to the specific nature of multilayer networks, are challenging methodological issues that, in our opinion, deserve future investigation.

- Also *cellular mobile wireless, ad-hoc wireless, and heterogeneous networks*—these are networks where an end-to-end connection may use different technological platforms and has to transverse several networks or routing domains with distinct technical features—are application environments that have and will be subject to significant evolutions that are and will be posing new, specific routing problems, where there are clearly new opportunities and challenges for the development of multi-criteria routing models in the near future. This is expected, keeping in mind the technical specificities of each network structure; the increasingly more complex nature of some of the routing problems (also reflecting the more complex nature of the network structures, specially in the case of heterogeneous networks); and the multi-dimensional characteristics, often conflicting, of the metrics and features that, desirably, are to be explicitly included in the routing models.

- Concerning *modeling aspects for the new technological platforms*, this will require, as far as routing models are concerned, the specification of adequate criteria, involving technical and often socioeconomic aspects, as illustrated in some of the references in Section 2.3.1. We also should mention the trend for including in network planning and design approaches, power consumption as a relevant criterion, not only by economic considerations but, foremost, also having in mind the already significant environmental impacts of energy consumption by telecommunication networks and information and communication technologies (ICT) structures in general. This concern also applies to routing models by considering the so-called "energy-aware routing methods," as illustrated in Wiatr et al. (2012) in which a heuristic routing method for WDM networks is presented that shows that there is a conflict between power consumption minimization and blocking probability. This is a specific area where adequate multi-criteria routing models should be developed, namely by extending previous models that did not include this criterion capable of exploring the trade-offs between power consumption and standard QoS and QoE objectives.

- We also should refer to issues that continue to be relevant, as a research topic, in the context of *QoS routing models that are explicitly multi-criteria*, keeping in mind the application of existing or of new multi-criteria routing methods dedicated to the new network technological platforms and, in particular, to heterogeneous or multilayer networks. A first issue to be addressed in a given application environment is obtaining better trade-offs in terms of exactness of the solution and computational efficiency. Note that many unicast routing models (without protection) of this type are formulated as multi-criteria shortest path models. This issue is particularly relevant in situations for which there is no feasible optimal solution, and the algorithm takes excessive time to detect such condition or to search for nondominated solutions in all areas of the objective function space of more interest (in terms of some system of preferences) or if the memory requirements are a practical constraint. This is often the case for networks of large dimension or connectivity, and this type of limitations is critically related to the so called "scalability" of the routing method that is the range of network dimensions or routing domains where the devised routing model can be applied. This is a critical concern that appears when we discuss a possible protocol implementation associated with a given multi-criteria algorithm.

- Regarding the complexity of exact algorithms, in our view, these should be the first type of approach, the applicability of which should be evaluated in this particular subarea of routing problems. We would like to remind that, although classical NP-completeness analysis is naturally relevant, this is a worst-case analysis, and, in many cases of application to routing methods, it may not be the decisive factor for choosing a resolution procedure. In fact, as noted in Kuipers and Mieghem (2005) worst-case complexity and execution times can be quite different in different cases of application. We think this is relevant, not only in "classical" QoS routing algorithms but also in some multi-criteria routing methods, such as the ones using exact multi-criteria shortest path formulations, tackled with efficient algorithms, and compatible with the required computational times, examples of which were referred to in Section 2.3.1. Similar considerations apply to some multicast routing models based on bi-criteria spanning tree algorithms, as illustrated in Craveirinha et al. (2013).

- Increasingly important as research themes, in relation to the new technological platforms and services, are routing methods that require the calculation of several paths simultaneously (i.e., *multipath routing models*). In particular, multicast routing involves the calculation of a set of paths from an originating node to multiple destination nodes, constituting a subset of the total node set, which involve the calculation of "minimum" (single criterion or multi-criteria)

Steiner trees. Of course these are research topics where new problems and challenges can be foreseen, taking into account, on the one hand the great complexity of the associated combinatorial problems and, on the other hand, the increasing multiplicity (and often the increasing structural complexity, as noted previously) of new technological platforms, network architectures, and service requirements. Topics of this kind would be, for example, the development of multi-criteria routing models for "anycast" flows associated with cloud computing (see Contreras et al. [2012]) or specific multicast routing methods, for example, for MPLS networks, where it is imposed that solutions include specific intermediate nodes, difficult problems that involve the obtainment of Steiner trees with special constraints. Note that related single-criterion unicast problems (intended for applications to resilient routing with path protection) concerning the calculation of shortest node disjoint path pairs visiting specific nodes was recently addressed through efficient heuristics in Gomes et al. (2017) and Martins et al. (2017).

- Another methodological trend is the development of *heuristics and metaheuristics* dedicated to the resolution of multi-criteria routing models in IP-based networks and multilayer networks, an area that is expected to continue growing in a near future. This has to do with various factors, now briefly revisited. Firstly, despite many classic NP-complete QoS routing problems having exact resolution methods, these may become intractable in networks of greater dimension or connectivity. This difficulty also may arise in many other cases, for example in models based on minimal spanning trees, where for larger dimensions of the networks, exact algorithms may become computationally too costly or even intractable. Secondly, in several cases, the introduction of new constraints may significantly complicate the base formulations. Thirdly, there are many other routing optimization problems that are NP-hard in the strong sense for which there are no exact resolution methods with execution times compatible with the applications. This is specially relevant in dynamic routing with very short routing update periods and in real-time routing. Another factor is the confluence in some routing models of one or several "complicating factors" in the sense described by Jones et al. (2002): very large number of variables (in particular in integer and mixed-integer formulations), non-linear objective functions/constraints, the explicit consideration of stochastic submodels in the problem formulation, and nonstandard utility functions as in many multi-criteria approaches (see, e.g., in Craveirinha et al. [2008]). These factors, articulated with the quite rapid increase in computing power as well as the advances in metaheuristic techniques and their availability in the Internet, have promoted the increasing importance of these approaches in the solution of many routing models as noted in Section 2.3. Nevertheless, we also would like to remark (as a result of our own analysis of various papers) that, in some cases, authors decide to use a priori heuristics or metaheuristics, ignoring the existence of exact approaches that could be applied to most of the practical configurations of the problem they are addressing.

- Finally, concerning the necessity of *evaluation and comparison of the performance* of routing models for a given network setting, we would like to remark that this is inherently a multi-criteria problem. In fact, multiple network performance metrics related to QoS and economic metrics (expected costs or revenues) should be included in such evaluation, whether a decision maker (typically a network engineer or a network manager) is seeking a preliminary evaluation or a final selection of a routing method. This is true whatever the types of routing methods, single-criterion, classical QoS, or multicriterion flow-oriented routing methods, and whenever more than one technical solution is available for a given network application. This is clearly, from an OR point of view, a problem involving classifying, ranking or selecting decision alternatives, according to multiple criteria/attributes, often conflicting and incommensurate, where the alternatives are in a small number and explicitly known, a priori. Note that this is particularly relevant for flow-oriented optimization routing approaches, taking into account their inherent limitations as analyzed in Craveirinha et al. (2008) and to single-criterion network-wide optimization routing methods. Take also note that the comparison of routing methods, in the vast literature in this area, has been based on empirical pair-wise comparison between methods. Naturally, it is adequate to tackle this issue by adequate multi-attribute decision analysis methods. A first, preliminary paper addressing this issue is Clímaco et al. (2015) where it is proposed the application of the variable interdependent parameter (VIP) methodology in Dias and Clímaco (2000) for a multi-attribute selection of flow-oriented routing methods, considering nine network performance attributes and taking into account the imprecise information associated with the relative importance of different network performance features. Also the treatment of this issue in a cooperative group decision setting is a research topic that should be addressed, taking into account that this type of decision, in reality, may involve more than one decision maker (e.g., two experts in network design with different technical opinions in specific aspects of the network metrics and a network manager).

2.4.2 Network Planning and Design

Regarding these two, quite interrelated areas, the following general topics can be explored:

- The study and development of new types of models (concerning new planning and design problems associated with the new technological platforms) associated with different decision processes.
- The development of new variants of the types of models presented in the outline of papers takes into account the implications on the planning processes of the fast technological evolution and its interaction with the turbulent transformations of the socioeconomic environment involving rapid market changes. Telecommunication applications with strong socioeconomic implications deserve further investment in multi-criteria modeling to enable a more realistic evaluation of their impacts. For instance, operational planning problems, vendor selection, e-commerce, and e-learning problems, etc.

It is expected that in the future some other problems in this area will be prone to treatment in a multi-criteria framework, especially with the very rapid and multifaceted technological evolutions previously identified (in their major aspects) and their interactions with complex and fast-changing economic and social trends. This trend is particular relevant to the new technological platforms, namely OTNs, 5G mobile wireless networks, and IoT.

Examples of such research challenges concern cell partitioning modes and frequency allocation problems involved in the complex planning and design process of mobile cellular networks. A third example in this area has to do with the design models of cooperative video streaming in which various network resources are to be pooled effectively by mobile video users in different application scenarios (see, e.g., Tang et al. [2017]) involving the optimization of various technical and economic factors.

Concerning the *modernization planning* of the access networks, the generalized introduction of broadband services (requiring optical fiber directly to the customer premises) is a type of problem in which different technological architectures can be used, so that a preliminary level of decision analysis for evaluating the alternatives seems worth considering. This level of analysis might be concerned with the evaluation, under different performance criteria (e.g., measuring upgrade cost, operator revenue, response to estimated demand, and user satisfaction in different technical instances) of various technologies and associated architectures available to the operator in a given market scenario, so the use of multicriteria models is clearly advantageous.

Concerning design issues of *wireless networks*, the reference Bourjolly et al. (2001) presents an overview of the application of OR-based decision support tools in this area. In particular the authors draw attention to the fact that cell partitioning (a decision process that uses several times the available frequencies and hence increases the network capacity) addresses two conflicting issues, namely covered area and capacity (involving, in essence, a choice between a smaller number of larger cells versus a larger number of smaller cells). As for the frequency allocation problem, it involves the assignment of a certain number of radio frequencies to each cell according to some "optimality" criteria and satisfying various technical constraints. In this type of problem, several objective functions can be considered, as discussed by those authors (i.e., the number of frequencies used, the frequency span, and two types of signal interference, all to be minimized). Now these models of wireless network design should also consider technical-economic and even environmental-related objectives, in particular power consumption, as illustrated in some references in Section 2.3.2. It will also be expected that new and complex problems of transmission design involving, in particular, the layout of mobile stations and the design of the associated antenna arrays, with multiple technical and economic objectives/constraints have been and will be fostered by the rapid expansion of mobile networks, WSNs, and ad hoc wireless networks. Also the possibility of choosing one or more different *suppliers of wireless communication services* of various sorts, in these different types of networks to be evaluated under multiple technical and economic instances, is an issue more and more topical. The evaluation of the *behavior of the subscribers* in face of the performance of the network operators and service suppliers is also a topical issue where, again, technical and socioeconomic criteria are at stake. The use of MCDA/M in this area, and in particular multiple-attribute decision analysis models, some recent examples of which were referred to in Section 2.3, is clearly another relevant recent trend.

A recent contribution in this context is Stocker and Whalley (2017) in which a multi-criteria model for the analysis of the broadband consumer experience is proposed, assuming that "speed isn't everything." Following these authors: "consumer experience may be affected by the increasingly complex nature of the value chain that provides online goods and services. In other words, faster broadband speeds do not necessarily result in an enhanced consumer experience." In fact, quality in Internet, contrary to common preconceptions, is only improved by increasing the speed until a certain level, depending on the case. In fact, its value can be represented by a U function. The overall quality involves "the perceived 'aggregate quality' that consumers experience when using a particular service of the Internet." This is evaluated through the QoE that integrates the technical QoS (this includes various quality of transmission and availability metrics), as we have seen previously, but it goes beyond the network/service performance characteristics, integrating also multiple aspects of the quality of the interaction with the network/service, as perceived by users (e.g., technical equipment available to the user, maintenance and billing, and consumer preferences). In conclusion, the authors show that the QoE evaluation is a difficult multi-criteria problem by discussing in some detail the dimensions under evaluation.

Of course, the next step is trying to use multi-criteria models to study this problem, being advised that it involves many imprecisions and uncertainties. It combines technical and economic criteria with very subjective ones.

These are areas which multi-criteria models, in particular multi-criteria location and capacity optimization problems (concerning, for example, the layout of mobile stations and the design of transmission systems in separate or in combination) and multi-attribute decision models (e.g., concerning the evaluation and choice of specific service suppliers in various network and market environments) constitute attractive/advantageous approaches. This is clearly the case, keeping in mind the tendency for the increased offer in specialized wireless services on a competitive basis, leading to a true "market of operators and services," an area where the development of multi-attribute approaches is clearly an important trend. Furthermore, several of the more complex of these decision problems, in particular those involving several, interrelated levels of decision/optimization, pose clear modeling and methodological challenges.

Heterogeneous networks, involving the interlacing of various technologies, namely concerning wired and wireless sections, is also an area in which the application of multi-criteria approaches has drawn increasing attention. A key problem in the design of such networks is the choice of a combination of transmission and signaling systems, seeking to achieve multiple objectives/requirements of QoS and of economic nature, some of which may be conflicting, a decision problem that clearly may be modeled as a multi-attribute decision model. This is naturally an area where the application of MCDA/M approaches should be investigated and where various decision problems should be tackled.

The design issues posed by *multilayered networks*, mentioned in the previous subsection, are multifaceted and usually involve multiple criteria. A primary issue is the analysis and comparison of different arquitectures. For example, when planning the development of optical fiber-based IP networks, various arquitectures can be used, namely, IP/MPLS over WDM, IP-over-OTN-over-WDM or MPLS-TP combined with OTN. Each of these configurations has specific technical features that may constitute an advantage or limitation concerning different capabilities, in particular with respect to wavelength switching, router capacities, power consumption, direct supporting of packet transport services, efficient bandwidth utilization, and resilience features; the economic features concerning capital expenditure (CAPEX), operational expenditure (OPEX), are also different, as are the cost features of different equipments to be used in such settings. The evaluation and choice of architectures of multilayered networks under various service demand and market scenarios is clearly a topic where multi-attribute analysis methods should be tested. The complex structures of these networks naturally have an impact in the OR formulations in general and in multi-criteria formulations of the design problems, in particular. Concerning the design problems, besides the interest in considering multi-criteria routing models, an issue previously addressed, the design of reliable telecommunication structures is a complex and challenging issue where multi-criteria approaches should be tested. The factors involved, namely resiliency objectives (i.e., the ability of the network to maintain an acceptable levels of QoS in the event of failures, namely equipment or software failures or abnormal working conditions, for example, unexpected overloads in parts of the network) and the economic evaluation of the network functioning in nominal and in failure conditions (typically aiming at the minimization of both CAPEX and OPEX) in multiple and uncertain scenarios, clearly open challenging fields for the development of MCDA/M approaches, from multi-criteria shortest paths applications to multiple-objective network flow-programming or multi-attribute decision analysis methods. An overview of research trends in the design of reliable telecommunication networks that can help in a better understanding of these issues can be seen in Rak et al. (2015).

Finally, we would like to emphasize the importance of modeling the uncertainty, in most of the problematic areas discussed so far, which requires particular attention in the future. In fact, the sources of uncertainty are multiple and of different natures.

2.4.3 Strategic Planning and Policy Issues

In this subsection we summarize the challenges raised by exciting strategical issues involving political, economic, social, and technological challenges. Hopefully, multi-criteria decision support systems will help in the clarification of options and consequences concerning difficult socio-technic-economic-political decisions.

A first class of problems is related to strategic telecommunication planning in countries and regions. Some papers on this issue were outlined in the previous section; however, they are mostly academic works. The real impact of these studies implies, in our view, several requisites: feeding the models with more and more accurate data; the involvement of the decision makers in the process so they can understand the advantages of using these tools; and the building of adequate/dedicate multi-criteria decision support systems. For instance, taking into account the difficulties associated with the use of many well-known methods in this context, the authors, following their recent experience, believe in the use of flexible learning oriented (open exchange) interactive tools because they seem to cope better with this type of problems. Furthermore, reinforcing the previous remarks, we believe that the following issues should be considered carefully: the involvement of several stakeholders/actors (cooperating or negotiating) in this area; the desirable public participation in situations where public and private spheres and the evolution of their borders are important issues; and the inevitability of coping with great uncertainties.

Secondly, structural implications of telecommunication networks evolutions in economy (growth/inequality of income distribution, employment issues, etc.), in society (media evolution, new services, etc.), and in governance (e-government, cybersecurity) also advise the use of multi-criteria evaluation. Although we do not know works tackling this issue and making explicit use of multi-criteria methods and decision support systems, the potential advantages of their use can be foreseen in some papers we will refer to. Next, a short outline of trends and challenges in this area is put forward.

As it is pointed by Jorgenson and Vu (2016), "[a] comprehensive ICT policy framework can be developed along seven dimensions: (i) connectivity and access; (ii) usage; (iii) legal and regulatory framework; (iv) production and trade; (v) skills and human resources; (vi) cybersecurity; and (vii) new applications. So, of course, ICT policy is a multi-dimensional effort involving conflicting issues; this justifies a strong effort to build adequate multi-criteria models to clarify the options of policy makers to minimize the dangers and potentiate their positive consequences, trying to improve and clarify the inevitable interactions among technological, economic, and political forces. On the one hand, new ICTs and, in particular, communication connectivity and new communication services, in general, potentiate growth increasing, but on the other hand, income distribution is creating large inequalities (i.e., a so called "digital divide"). Ogunsola (2005) discusses this issue, using the enlightening concept of "digital slavery." Bauer (2014, 2018) introduces, in a very clear manner, the strong connections among technological change (wireless communication and new services; fusion of computing and connectivity, changing the dynamics of competition; fully algorithmicized platform markets; machine learning, AI; robotics, Internet of Things, etc.) and productivity, as well as the creation of new relations between location of production and workforce. As it is recognized in Bauer (2014), "neither the theory of platform markets nor the ecosystem approach has yet resulted in a set of practical guidelines that could be implemented without further work by the regulatory agencies." As in many other periods of the history of mankind, the rapid progress of science and technology opens incredible possibilities for improving human life in multiple spheres but also generates dangers and challenges. However, the present situation is peculiar because the new ICT technologies potentiated the globalization of the world activities. In these circumstances, although a bright future is theoretically possible, we believe that if the present trend of capitalism is not changed very fast by regulations and other policies, in the near future we may be faced with a destructive crisis. The speed of technological changes is incredible, and the present "rules of the game" (in particular, the short-term objective of maximizing the profits of company shareholders) are leading to large and always increasing income inequalities. Regarding the employment trends, the present situation is also unsustainable. In Bauer (2018) concluded that "these developments have contributed to wage polarization, the growth of high-paying and unskilled low-payment jobs, while the number of jobs requiring a middle level of education and paying middle wages is shrinking." We recognize this is the major trend in the developed world, although the exclusion is remarkable even in this part of the world. Two digital divides are being created, one inside the rich countries and a second one between these countries and the majority of the world, still underdeveloped. Of course, the defenders of the present "rules of the game" say that humanity never had such good conditions, even the poor, and that, in the long term, if everything is deregulated, the markets will correct the distortions. The second statement is just a belief impossible to be supported scientifically, empirically, or rationally. Furthermore, as Keynes pointed out (Keynes, 1923), in the long term, we are all dead, so short- and medium-term expectations are essential for present generations. On the other hand, the first statement is true. However, the ways of living in different stages of evolution of human societies are neither fair nor adequate the direct comparison.

Fortunately, globalization enables poor people the world over to be aware of the state of living in the different parts of the world. It is our conviction that, in our own interest, "rules of the game" should be changed through regulation and other policies. Otherwise, sooner or later, big social tensions and fragmentations will lead to disastrous convulsions.

According to Bauer (2018), "whether the inequality-increasing or inequality-decreasing forces dominate, depends on economic, technological, and institutional context in which they unfold as well as on the presence of mitigating policy measures."

Moreover, Jorgenson and Vu (2016) suggest that "[p]riority should be given to e-government, e-business and internet enabled services." Recognizing their great potential increasing productivity, enabling a better resource allocation, improving quality of life of citizens, etc., we must not forget the associated dangers. Besides some of the aspects previously referred to, cyber security and privacy intrusion issues must be emphasized, especially in relation to the widespread of cloud computing and the expected exponential growth of the IoT, concerns that should also be taken into account in multi-criteria approaches focusing on some strategic modernization planning problems.

In all these situations we believe that interactive multi-criteria decision support systems, in many cases rooted in learning-oriented tools, can be useful, more than for decision-aiding support to help the clarification of decision options and respective consequences for the involved actors, including the intervention of citizens individually or organized. This is a huge challenge. In fact, such tools do not exist yet.

As we have seen, mathematical programming and multi-attribute models have been used, depending on the case. In some situations, we believe in the complimentary use of both types of approaches. Mathematical programming approaches can be useful to evaluate, in a first step, the feasible alternatives, usually in big number, using a restricted number of criteria and satisfying some other restrictions, to make the most of their computational usefulness. This first step would enable the reduction of the scope of the search, helping in the identification of a limited number of alternatives requiring deeper evaluation, which would be done in the second stage. In this second step, multi-attribute models provide a deeper analysis of the problem under study by evaluating a small number of alternatives, in a more detailed or extensive family of criteria.

Thirdly, we outline the challenges and future trends regarding the strategic planning issues concerning the assignment and allocation of the broadband spectrum. Once more, there are strong conflictual challenges, namely opposing efficiency and equity issues. First of all, it must be remarked we are living a transition from an administrative assignment of bands to a new deregulation based on auctions.

In Minervini (2014), different strategies and tactics for deregulation of the spectrum use are discussed. The author concludes that "a gradual approach is preferable if it offers options to deal with uncertainty better, by acting on reform sequencing to reduce uncertainty and to maximize expected payoff." This is a first point where multi-criteria models are potentially useful. Cave and Nichols (2017) observe that "as demand for mobile communications grew, spectrum in additional bands was auctioned, operators built up portfolios, and their market share diverges. The old system of assigning to the highest bidders a chosen number of licences of equal and predetermined size in a single band gave way to multiband auctions in which the auction process itself determined the size of the award by each operator." These authors consider that multiple objectives, efficiency and equity objectives, should be attained. In the future, it is foreseeable the use of multi-criteria models to clarify the auction design, combining equity and efficiency objectives.

Nevertheless, we believe that, in this area, a strong regulatory and supervisory intervention, namely by national or supranational authorities, should be preserved. Such intervention should have in mind, on the one hand, the inherent public nature of the radio-frequency spectrum (a natural physical resource, unlike other transmission media) and, on the other hand, the need to guarantee technical coordination and, foremost, a balanced working of such specific "market," avoiding explicit or tacit monopolistic situations.

Finally, we refer to two papers dealing with important issues in this area, where the introduction of multi-criteria analysis may also be justified. The first deals with the problem of the externalities associated with broadband frequency assignment. Cave and Pratt (2016) deal with the important issue of taking into account externalities when spectrum is assigned and allocated to broadcasting and mobile communications. Planning the broadband use implies the evaluation of public values of alternative uses, involving social, economic, financial, and political issues. Of course, taking into account externalities can make the difference, however in many situations, neither the firms nor the users have market-direct relations and so monetizing their values is not easy. In these circumstances, one should discuss how to measure and valuing externalities, as tackled in Cave and Pratt (2016). The second issue is related to the spectrum-sharing potentialities and risks. Cui et al. (2017) discuss the multi-dimensional issue of compromising among the efficiency, flexibility, and some QoS advantages of sharing spectrum with the associated risks. Of course, the use of multi-criteria models incorporating uncertainties and risks is clearly useful in this context.

ACKNOWLEDGMENTS

This work was partially supported by FCT ("Fundação de Ciência e Tecnologia") under project grant UID/Multi/00308/2019 and by FEDER Funds and National Funds through FCT under project CENTRO-01-0145-FEDER-029312.

REFERENCES

Abourezq, M., A. Idrissi (2015). Integration of QoS aspects in the cloud computing research and selection system, *International Journal of Advanced Computer Science and Application (IJACSA)*, 6(6), 1–13.

Adebiyi, S., E. Oyatoye, O. Kuye (2015). An analytic hierarchy process analysis: Application to subscriber retention decisions in the Nigerian mobile telecommunications, *International Journal of Management and Economics*, 48, 63–83.

Aissanou, F., A. Petrowski (2013). Autonomous multi-criteria decision making for route selection in a telecommunication network, *Proceedings 2013 IEEE Symposium on Computational Intelligence in Multi-Criteria Decision-Making (MCDM)*, pp. 33–40, Singapore.

Ali, A., W. Hamouda, M. Uysal (2015). Next generation M2M cellular networks: Challenges and practical considerations, *IEEE Communications Magazine*, 53(9), 18–24.

Awduche, D., L. Berger, D. Gan, T. Li, V. Srinivasan, G. Swallow (2001). RSVP-TE: Extensions to RSVP for LSP tunnels. *IETF RFC 3209.*

Bauer, J.M. (2014). Platforms, systems competition, and innovation: Reassessing the foundations of communications policy, *Telecommunications Policy*, 38, 662–673.

Bauer, J.M. (2018). The internet and income inequality: Socio-economic challenges in a hyperconnected society, *Telecommunications Policy*, 42(4), pp. 333–343.

Bhat, S., G. Rouskas (2016). On routing algorithms for open marketplaces of path services. *2016 IEEE International Conference on Communications (ICC)*, Kuala Lumpur, Malaysia.

Bentes, A.V., J. Carneiro, J.F. da Silva, H. Kimura (2012). Multidimensional assessment of organizational performance: Integrating BSC and AHP. *Journal of Business Research*, 65(12), 1790–1799.

Bezruk, V., A. Bukhanko, D. Chebotaryova, V. Varich (2012). Multi-criteria optimization in telecommunication networks planning, designing and controlling, in Dr. Jesús Ortiz (Ed.) *Telecommunications Networks – Current Status and Future Trends*, InTech, pp. 252–274. ISBN: 978-953-51-0341-7.

Bhunia, S., S. Roy, N. Mukherjee (2014). Adaptive learning assisted routing in wireless sensor network using multi criteria decision model. *International Conference on Advances in Computing, Communications and Informatics (ICACCI, 2014)*, New Delhi, India, pp. 2149–2154.

Bourjolly, J., L. Déjoie, K. Dioume, M. Lominy (2001). Frequency allocation in cellular phone networks: An OR success story, *OR/MS Today*, 28(2), 41–44.

Bouyssou, D. (1990). Building criteria: A prerequisite for MCDA, in C.A. Bana e Costa (Ed.) *Readings in Multiple Criteria Decision Aid*, Springer Verlag, Berlin, Germany, pp. 58–80.

Brans, J., B. Mareschal (2005). Promethee methods, in J. Figuiera, S. Greco, and M. Ehrgott (Eds.) *White Space Communications*, Multiple Criteria Decision Analysis: State of the Art Surveys, Springer, Dordrecht, the Netherlands, pp. 163–196.

Bueno, M.L.P., G.M.B. Oliveira (2014). Four-objective formulations of multicast flows via evolutionary algorithms with quality demands, *Telecommunication Systems*, 55(3), 435–448.

Cave, M., N. Pratt (2016). Taking account of service externalities when spectrum is allocated and assigned, *Telecommunications Policy*, 40, 971–981.

Cave, M., R. Nicholls (2017). The use of spectrum auctions to attain multiple objectives: Policy implications, *Telecommunications Policy*, 41, 367–378.

Çetinkaya, E.K., J.P.G. Sterbenz (2013). A taxonomy of network challenges. *In Proceedings of the 9th IEEE/IFIP International Conference on Design of Reliable Communication Networks (DRCN)*, Budapest, Hungary, pp. 322–330.

Chen, C.-T., W.-Z. Hung (2009). Applying ELECTRE and maximizing deviation method for stock portfolio selection under fuzzy environment, *Studies in Computational Intelligence*, 214, 85–91.

Chen, M., H. Bai, Y. Zhou, Z. Wang, P. Jiang (2014). A novel network performance evaluation method based on maximizing deviations, *Telecommunication Systems*, 55, 149. doi:10.1007/s11235-013-9759-1.

Cisco VNI, Global Data Traffic Forecast Update, 2015–2020, white paper Dec. 2016.

Cisco, Visual Networking Index: Global Mobile Data Traffic Forecast Update, 2016–2021 White Paper, March 28, 2017, Document ID:1454457600805266, https://www.cisco.com/c/en/us/solutions/collateral/service-provider/visual-networking-index-vni/mobile-white-paper-c11-520862.html.

Clímaco, J., M.E. Captivo, M. Pascoal (2010). On the bicriterion—Minimal cost/minimal label—Spanning tree problem, *European Journal of Operational Research*, 204(2), 199–205.

Clímaco, J., M. Pascoal (2012). Multicriteria path and tree problems: Discussion on exact algorithms and applications, *International Transactions in Operational Research*, 19(1–2), 63–98.

Clímaco, J., J. Craveirinha, L. Martins (2015). Cooperative group multi-attribute analysis of routing models for telecommunication networks, *Proceedings Conference Group Decision and Negotiation - GDN 2015*, B. Kaminski, G. Kersten, P. Szufel, M. Jakubczyk, T. Wachowicz (Eds.) Warsaw School of Economics Press, Warsaw, Poland, pp. 177–184, 2015.

Clímaco, J., J. Craveirinha, R. Girão-Silva (2016). Multicriteria analysis in telecommunication network planning and design—A survey, in S. Greco, M. Ehrgott, J. Figueira (Eds.) *Multiple Criteria Decision Analysis - State of the Art Surveys*, International Series in Operations Research & Management Science, vol. 233, Chapter 26, pp. 1167–1233, Springer.

Colson, G., K. Sabri, M. Mbangala (2006). Multiple criteria and multiple periods performance analysis: The comparison of telecommunications sectors in the Maghreb countries, *Journal of Telecommunications and Information Technology*, 4, 67–80.

Contreras, L.M., V. Lopez, O.G. De Dios, A. Tovar, F. Munoz, A. Azañón, J.P. Fernandez-Palacios, J. Folgueira (2012). Toward cloud-ready transport networks, *IEEE Communications Magazine*, 50(9), 48–55.

Craveirinha, J., R. Girão-Silva, J.A. Clímaco (2008). A meta-model for multiobjective routing in MPLS networks, *Central European Journal of Operations Research*, 16(1), 79–105.

Craveirinha, J., J. Clímaco, L. Martins, C.G. Silva, N. Ferreira (2013). A bi-criteria minimum spanning tree routing model for MPLS/overlay networks, *Telecommunication Systems*, 52(1), 203–215.

Craveirinha, J., J. Clímaco, L. Martins, M. Pascoal (2016). An exact method for constructing minimal cost/minimal SRLG spanning trees over optical networks, *Telecommunication Systems*, 62(2), 327–346.

CTIA, The US Wireless Association, U.S. wireless—Quick facts, 2017, https://www.ctia.org/industry-data/facts.

Cui, L., M.B.H. Weiss, B. Morel, D. Tipper (2017). Risk and decision analysis of dynamic spectrum access, *Telecommunications Policy*, 41, 405–421.

Das, B., S. Bhunia, S. Roy, N. Mukherjee (2015). Multi-criteria routing in wireless sensor network using weighted product model and relative rating, *Proceedings Applications and Innovations in Mobile Computing (AIMoC)*, Kolkata, India.

Desruelle, P., J. Stancik (2014). Characterizing and comparing the evolution of the major global economies in information and communication technologies, *Telecommunications Policy*, 38, 812–826.

Dias, L., J. Clímaco (2000). Additive aggregation with interdependent parameters: The VIP analysis software, *Journal of the Operational Research Society*, 51, 1070–1082.

El-Sayed, M., J. Jaffe (2002). A view of telecommunications network evolution, *IEEE Communications Magazine*, 40(12), 74–81.

Esteves, J.S., J. Craveirinha (2013). On a bicriterion server allocation problem in a multidimensional erlang loss system, *Journal of Computational and Applied Mathematics*, 252, 103–119.

Figueira, J., V. Mousseau, B. Roy (2016). ELECTRE methods, chapt in multiple criteria decision analysis–State of the art surveys. In S. Greco, M. Ehrgott, J. Figueira (Eds.) *International Series in Operations Research & Management Science*, vol. 1, Part III, Chapter 5, Springer+Business Media, New York.

Findeisen, W., F. Bailey, M. Brdys, K. Malinkowski, P. Tatjewski, A. Wozniak (1980). *Control and Coordination of Hierarchical Systems*, John Wiley & Sons, Chichester, UK.

Gallis, A., S. Clayman, L. Mamatas, J.R. Loyola, A. Manzalini, S. Kuklinski, J. Serrat, T. Zahariadis (2013). Softwarization of future networks and services programmable enabled networks as next generation defined networks, *IEE Workshop SDN for Future Networks and Services*, Trento, Italy.

Gerpott, T.J., N. Ahmadi (2015). Advancement of indices assessing a nation's telecommunications development status: A PLS structural equation analysis of over 100 countries, *Telecommunications Policy*, 39, 93–11.

Girão-Silva, R., J. Craveirinha, J. Clímaco (2012). Hierarchical multiobjective routing model in multiprotocol label switching networks with two service classes—A pareto archive strategy, *Engineering Optimization*, 44(5), 613–635.

Girão-Silva, R., J. Craveirinha, J. Clímaco, M.E. Captivo (2015). Multiobjective routing in multiservice MPLS networks with traffic splitting—A network flow approach, *Journal of Systems Science and Systems Engineering*, 24(4), 389–432.

Girão-Silva, R., J. Craveirinha, T. Gomes, L. Martins, J. Clímaco, J. Campos (2017). A network-wide exact optimization approach for multiobjective routing with path protection in multiservice multiprotocol label switching networks, *Journal of Engineering Optimization*, 49(7), 1226–1246.

Gomes, T., J. Silva, J. Craveirinha, C. Simões (2012). Protected bicriteria paths in transport networks, in J. Rak, M. Pickavet, H. Yoshino (Eds.) *RNDM 2012, Proceedings da 4th International Workshop on Reliable Networks Design and Modeling*, pp. 91–97, St. Petersburg, Rússia.

Gomes, T., L. Jorge, P. Melo, R. Girão-Silva (2016). Maximally node and SRLG-disjoint path pair of min-sum cost in GMPLS networks: A lexicographic approach, *Photonic Network Communications* 31, 11–22.

Gomes, T., L. Martins, S. Ferreira, M. Pascoal, D. Tipper (2017). Algorithms for determining a node-disjoint path pair visiting specified nodes, *Optical Switching and Networking*, 23, 189–204.

Gomes da Silva, C., J. Clímaco (2007). A note on the computation of ordered supported non-dominated solutions in bicriteria minimum spanning tree problems, *Journal of Telecommunications and Information Technology*, 4, 11–15.

González, D., M. García-Lozano, S. Ruiz, M.A. Lema, D. Lee (2016). Multiobjective optimization of fractional frequency reuse for irregular OFDMA macrocellular deployments, *Telecommunication Systems*, 61, 659. doi:10.1007/s11235-015-0060-3.

Gouveia, L., P. Patrício, A. de Sousa (2016). Lexicographical minimization of routing hops in hop-constrained node survivable networks, *Telecommunication Systems*, 62, 417–434.

Grzegorek, J., A.P. Wierzbicki (2012). Multiple criteria evaluation and ranking of social penetration of information society technologies, *Journal of Telecommunications and Information Technology*, 4, 3–13.

Handley, M. (2006). Why the internet only just works, *British Telecom Technology Journal*, 24(3), 119–129.

Hwang, C.L., K. Yoon (1981). *Multiple Attribute Decision Making Methods and Applications*, Springer Verlag, New York.

ITU-T, Network node interface for the optical transport network (OTN). Rec. G.709/Y.1331 (2009).

Jones, D.F., S.K. Mirrazavi, M. Tamiz (2002). Multiobjective metaheuristics: An overview of the current state-of-the-art, *European Journal of Operational Research*, 137(1), 1–9.

Jorgenson, D.W., K.M. Vu (2016). The ICT revolution, world economic growth, and policy issues, *Telecommunications Policy*, 40, 383–397.

Kaplan, R., D. Norton (1996). *The Balanced Scorecard Translating Strategy into Action*, Harvard Business School Press, Boston, MA.

Keeney, R.L. (2001). Modeling values for telecommunications management, *IEEE Transactions on Engineering Management*, 48(3), 370–379.

Keeney, R.L., H. Raiffa (1993). *Decisions with Multiple Objectives: Preferences and Value Tradeoffs*, Cambridge University Press.

Keynes, J.M. (1923). *A Tract on Monetary Reform*, Chapter 3, p. 80, Macmillan and Co, London, UK.

KU TopView ResiliNets Topology Map Viewer, 2011, 21 Nov 2012, http://www.ittc.ku.edu/resilinets/maps/

Kuipers, F.A., P.F. Van Mieghem (2005). Conditions that impact the complexity of QoS routing, *IEEE/ACM Transactions on Networking*, 13(4), 717–730.

Lee, W.C., M.G. Hluchyj, P.A. Humblet (1995). Routing subject to quality of service constraints in integrated communication networks, *IEEE Networks*, 9(4), 46–55.

Martins, L., T. Gomes, D. Tipper (2017). Efficient heuristics for determining node-disjoint path pairs visiting specified nodes, *Networks*, 70(4), 292–307.

Messac, A., A. Ismail-Yahaya, C.A. Mattson (2003). The normalized normal constraint method for generating the pareto frontier, *Structural and Multidisciplinary Optimization*, 25(2), 86–98.

Mfupe, L., F. Mekuria, M. Mzyece (2017). Multicriteria decision analysis of spectrum management frameworks for futuristic wireless networks: The context of developing countries, *Mobile Information Systems*, 2017, Article ID 8610353 (18 pages), 1–18.

Minervini, L.F. (2014). Spectrum management reform: Rethinking practices, *Telecommunications Policy*, 38, 136–146.

Mohanty, S., D. Dabade (2015). Vendor selection for service sector industry: A case study on supplier selection to Indian telecom service provider using AHP technique, *IOSR Journal of Business and Management (IOSR-JBM)*, 2319-7668, 32–44.

Monserrat, J. et al. (2014). Rethinking the mobile and wireless network architecture: The METIS research into 5G, *Proceedings European Conference Network and Communication*, 1–5.

Niven-Jenkins, B., D. Brungard, M. Betts, N. Sprecher, S. Ueno (2009). Requirements of an MPLS transport profile. *IETF RFC 5654*.

Nurminen, J.K. (2003). Models and algorithms for network planning tools-Practical experiences, *Research Report E14, Systems Analysis Laboratory*, Helsinki University of Technology.

Ogunsola, L.A. (2005). Information and communication technologies and the effects of globalisation: Twenty-first century digital slavery for developing countries—Myth or reality, *Electronic Journal of Academic and Special Librarianship*, 6, 1–2.

Pereira, R.A., D. Bianchini (2013). Application of method AHP in the decision for reduction of the levels of legal action in companies of telecommunications, *8º CONTECSI—International Conference on Information Systems and Technology Management*, pp. 2878–2902.

Perera, C., C.H. Liu, S. Jayawardena, M. Chen (2014). A survey of Internet of Things: From industrial market perspective, *IEEE Access*, 2, 1660–1679.

Rak, J. et al. (2015). Future research directions in design of reliable communication systems, *Telecommunication Systems*, 60(4), 423–450.

Rehena, Z., S. Roy, N.I. Mukherjee (2017). Multi-criteria routing in a partitioned wireless sensor network, *Wireless Personal Communications*, 94(4), 3415–3449.

Reid, A., P. Willis, I. Hawkins, C. Bilton (2008). Carrier ethernet, *IEEE Communications Magazine*, 46(9), 96–103.

Rostami, A., P. Ohlen, K. Wang, Z. Ghebretensae, B. Skubic, M. Santos, A. Vidal (2017). Orchestration of RAN and transport networks for 5G: An SDN approach, *IEEE Communications Magazine*, April 2017, pp. 64–70.

Roy, B., D. Bouyssou (1993). *Aide Multicritére à la Deécision: Méthodes et Cas*. Economica, Paris, France.

Saaty, T.L. (1980). *The Analytic Hierarchy Process: Planning, Priority Setting, Resource Allocation. Decision Making*, McGraw-Hill, New York.

Saaty, T.L. (1994a). Highlights and critical points in the theory and application of the analytic hierarchy process, *European Journal of Operational Research*, 74(3), 426–447.

Saaty, T.L. (1994b). How to make a decision: The analytic hierarchy process, *Interfaces*, 24(6), 19–43.

Sahli, N., N. Jabeur, I.M. Khan, M. Badra (2012). Towards a generic framework for wireless sensor network multi-criteria routing, *2012 5th International Conference on New Technologies, Mobility and Security (NTMS)*, Istanbul, Turkey. doi:10.1109/NTMS.2012.6208737.

Savitha, K., C. Chandrasekar (2011).Vertical handover decision schemes using SAW and WPM for network selection in heterogeneous wireless networks, *Global Journal of Computer Science and Technology*, 11(9), 19–24.

Shi, F., W. Liu, D. Jin, D. Weijie, J. Song (2014). A cluster-based countermeasure against blackhole attacks in MANETs, *Telecommunication Systems*, 57, 119–136.

Slowinski, R., S. Greco, B. Matarazzo (2012). Rough set and rule-based multicriteria decision aiding, *Pesquisa Operacional*, 32(2), 213–270.

Steuer, R.E. (1986). *Multiple Criteria Optimization: Theory, Computation and Application. Probability and Mathematical Statistics*, John Wiley & Sons, Hoboken, NJ.

Statnikov, R., J. Matusov, K. Pyankov, A. Statnikov (2013). Multi-criteria optimization of cellular networks, *Open Journal of Optimization*, 53–60. doi:10.4236/ojop.2013.23008.

Statnikov, R., A. Statnikov (2011). *The Parameter Space Investigation Method Toolkit*, Artech House, Boston, MA.

Stocker, V., J. Whalley (2017). Speed isn't everything: A multi-criteria analysis of the broadband consumer experience in the UK, *Telecommunications Policy*, 42, 1–14.

Suh, Y., K.T. Kim, D.R. Shin, H.Y. Youn (2015). Traffic-aware energy efficient routing (TEER) using multi-criteria decision making for wireless sensor network. *5th International Conference on IT Convergence and Security (ICITCS)*, Kuala Lumpur, Malaysia.

Tang, M., L. Gao, H. Pang, J. Huang, L. Sun (2017). Optimizations and economics of crowdsourced mobile streaming, *IEEE Communications Magazine*, 55(4), 21–27.

Tran, T., A.Hajsami, P.Pandey, D. Pompili (2017). Collaborative mobile edge computing in 5G networks: New paradigms, scenarios and challenges, *IEEE Communications Magazine*, pp. 54–63.

Uygun, Ö., H. Kaçamak, Ü. A. Kahraman (2015). An integrated DEMATEL and fuzzy ANP techniques for evaluation and selection of outsourcing provider for a telecommunication company, *Computers & Industrial Engineering*, 86, 137–146. doi:10.1016/j.cie.2014.09.014.

Wiatr, P., P. Monti, L. Wosinska (2012). Power savings versus network performance in dynamically provisioned WDM networks. *IEEE Communications Magazine*, 50(5), 48–55.

Wierzbicki, A. P., W. Burakowski (2011). A conceptual framework for multiple-criteria routing in QoS IP networks, *International Transactions in Operational Research*, 18(3), 377–399.

Wojewnik, P., T. Szapiro (2010). Bi-reference procedure BIP for interactive multicriteria optimization with fuzzy coefficients, *Central European Journal of Economic Modelling and Econometrics*, 2(3), 169–193.

Zanella, A., N. Bui, A. Castellani, L. Vangelista, M. Zorzi (2014). Internet of Things for smart cities, *IEEE, Internet of Things Journal*, 1, 22–32.

Zitzler, E., M. Laumanns, L. Thiele (2002). SPEA2: Improving the strength pareto evolutionary algorithm for multiobjective optimization, in Evolutionary Methods for Design, Optimisation and Control with Application to Industrial Problems (EUROGEN 2001).

CHAPTER 3

SISTI

A Multicriteria Approach to Structure Complex Decision Problems

Maria Franca Norese

CONTENTS

3.1 INTRODUCTION

A multi-criteria decision-aid (MCDA) process should be the result of interactions among analysts, decision makers, and stakeholders, but decision aiding is sometimes required when a problem situation is new and a formal decision system with well-defined rules, clear constraints, roles and relations of the actors does not exist. This situation arises frequently when the need for the activation of innovative situations is perceived, recognized, and proposed by actors (people or organizations) in connected decision processes, who need to study the specific nature of the problem situation before the activation of a new decision process, or by experts in a specific domain, in relation to a new and badly structured research theme.

Formal or informal documents may be available, and they may be used to understand the organizational context and define the decision problem. When structured knowledge and data are not available, the need for a course of action in relation to the new and not sufficiently defined problem generates a request for investigation, data acquisition, and elaboration. However, these activities are often not clearly defined and not aimed at a specific goal because of a lack of knowledge and specific competences. Moreover, their development and results cannot be oriented and controlled because decision authority and accountability have not yet been foreseen. When data and possible indicators are easily accessible in institutional databases, their use in new and ill-structured situations is often characterized by a high multiplicity of items or indicators because of the general belief that only a large amount of data can produce information. An integration of these data becomes difficult and, therefore, risky for at least two reasons: first because a logical structure of the problem and its information needs had not been generated before and also because a synthesis of such different and "incomparable" elements, from different and often inaccessible sources, is not easy.

Adopting Simon's three-phase framework of Intelligence, Design, and Choice (Simon, 1960, 1991), an MCDA process can also be developed in these situations and oriented toward facilitating the Intelligence phase of a not-yet institutional decision process, which requires understanding and problem structuring. However, some precautions as well as focused and full attention to the structuring, modeling, and control activities are required.

MCDA adopts a *constructivist approach* in which a model, concepts, and procedures are not envisaged to reflect a well-defined reality that exists independently of the actors but as a communication tool (Tsoukias, 2007) that allows the participants in the decision process to carry forward a process of thinking and to talk about the problem (Genard and Pirlot, 2002). A constructivist approach cannot be applied in situations in which only some actors perceive the nature and importance of the decision problem and in which there are not sufficient conditions to activate a decision process and a decision system. However, an effective interaction with the few potentially involved actors and a preliminary study, which should include problem structuring, multi-criteria (MC) modeling, application of

MC methods, and result analysis and validation, become useful to clarify a complex and new situation, reduce uncertainty, and structure the relevant complexity elements in a "good" model of the problem situation.

For this reason, a new methodological approach has been proposed and is here called SISTI, to operate in relation to new, unstructured and complex decision problems, in which the decision makers' involvement is limited or even absent. This kind of intervention is a *SImulated* decision aiding intervention because the decision process and the system are not yet activated and a decision-aid process is not oriented toward an immediate decision. Moreover, it may be described as a *STImulating* approach because the study is developed together with the few actors that perceive the need to understand and propose structured and consistent (and then stimulating) elements for later phases of a still-not activated decision process.

Understanding the problem situation and structuring its different aspects is essential in the presence of limited knowledge of the problem and informative uncertainty. Literature proposes the activation of problem structuring methods (see Rosenhead, 1996, for the definition and Rosenhead and Mingers, 2001, for the theory and practice), to orient an analysis, to create a common interpretation and understanding space, to structure the analyzed system, and to elaborate a clear problem formulation. The aims are similar in the *description problem statement*, an approach that was introduced in Roy (1981) and described in Roy (1985, 1996) as *a rudimentary type of aid that poses rather than solves the problem*. This approach aims to clarify a decision situation by identifying and describing the elements of an MC model without making any recommendation. When the decision maker's active part in the decision-aid process is limited or impossible, this approach can be facilitated by introducing an *open model*, as described in Vanderpooten (2002), that is, a framework that supports reflection, investigation, or negotiation.

SISTI makes this framework operational in a descriptive problem statement, without the proposal of a new MC method. SISTI has been proposed to improve understanding, reduce uncertainty, and then facilitate problem structuring and modeling using a multicriteria approach (which connects the possibility of using MC methods with a specific way of acting on representations of a different nature and formalization level, as described in de Montgolfier and Bertier [1978]) and to propose the logic and terminology of MC methods from the beginning of an intervention. This methodological approach has been stimulated by the requests of decision aiding in these difficult decision situations but also by the purpose of improving the modeling ability of students, young researchers, and practitioners.

The main factors that characterize SISTI are described in the next section with details of the activities that are required to implement the approach and the description of the SISTI evolution from an unusual intervention to a methodology. In the Section 3.3, SISTI is analyzed in relation to a specific application in which visualization and comparison of different results are essential elements. The potentialities and drawbacks of SISTI and the need to make SISTI less demanding in practice are synthesized in the conclusions.

3.2 THE SISTI APPROACH TO STRUCTURE A DECISION PROBLEM

A new problem situation in a given organizational context requires a perspective and a specifically orientated decision-aid process to identify the main complexity and uncertainty factors and then to use this knowledge to control uncertainty and make complexity explicit and "manageable" by means of structured concepts, models, and consistent methods.

Sometimes an unstructured problem and complex decision context make it necessary to combine a problem structuring method and an MC method as described in Belton et al. (1997), Bana e Costa (1999), Rosenhead and Mingers (2001), and more recently in Petkov et al. (2007), Montibeller et al. (2008), Norese et al. (2008), Stewart et al. (2010), and Belton and Stewart (2010). A drawback of this interesting multi-methodology approach could be the need to introduce two different methodological approaches in a real and "messy" decision context. SISTI can instead be used to face these situations by means of the introduction of only one methodological approach and its formal language, both to structure the problem and to use MC methods to recommend possible solutions or conclusions.

In general, an MCDA process is developed in interaction with decision makers and stakeholders. However, when the problem is new and not sufficiently structured, their roles are often not yet clearly defined. If the problem does not require an immediate decision, the main aim of a decision-aid intervention may be the understanding of the situation, and SISTI is a methodological approach that can be proposed for this kind of MCDA intervention. When an intervention starts, a realistic representation of the problem at hand is ill structured or unstructured. The original problem formulation gradually evolves in the course of the process and some aspects may change importance level or have to be determined and consolidated. The advantage of using the MCDA framework as a structuring template to guide the representation of the "mess" as a problem defined in terms of criteria, alternatives, uncertainties, stakeholders and environment "is that the transition from divergent structuring to convergent analysis is essentially seamless" (Belton and Stewart, 2010).

MC models are used in SISTI above all because:

- They are able to transparently include all the relevant aspects of a decision problem using the descriptive and procedural terminology of each specific analysis field.
- They are structured with the aim of eliminating redundancies, including the minimal set of essential and consistent elements and distinguishing between data and reliable evaluations.

SISTI is a model-based process that is activated to develop a sequence of temporary MC models oriented toward clarifying the situation, reducing uncertainty, converging toward a robust model that is consistent with the problem situation, and proposing an effective approach for the later phases of a decision process.

SISTI includes a recurring cycle of different activities. At the beginning, a draft MC model, which may be taken from literature, such as the expression of the experts in the involved fields, or may be set up directly on the basis of the examined context, is employed only as a basis of reasoning to represent some first conceptual hypotheses in formal terms and to activate the first SISTI cycle. An MC method and a critical analysis of the result of its application to a model are used here, at each iteration, to identify any weakness elements of the result and, therefore, of the model or of the modeling hypothesis that has led to the generation of the model. A sensitivity analysis (which concentrates attention on how the result changes in relation to the values of each single parameter) is developed in relation to the identified characteristics of the result. A structured synthesis of these results is used to verify whether a marginal or structural change in the (formal or conceptual) model would be possible and could generate a useful improvement.

This sequence of activities (identification of the main characteristics of the result, development of a sensitivity analysis, and translation of its result into an easily understandable framework that can be used in a communication context) can be defined a result analysis (RA) procedure (see Figure 3.1). RA is developed at a technical level, but its temporary results

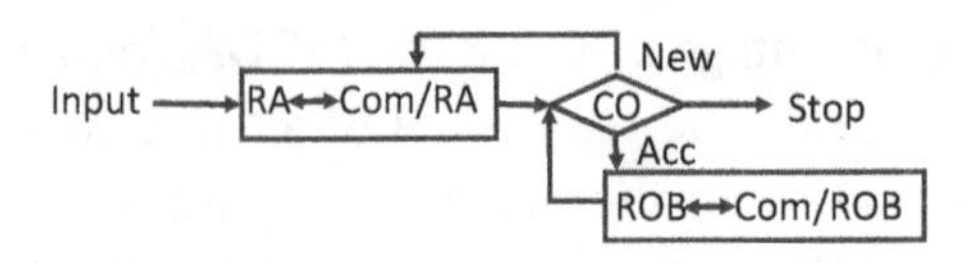

FIGURE 3.1 Logical sequence of the SISTI activities.

have to be analyzed with the few actors involved in the various steps of the SISTI process (Com/RA). A collective visualization of some scenarios, which concentrates attention on how the result changes in relation to the values of some parameters, and the examination of possible weakness elements of a result produce new knowledge and may include elements that stimulate a marginal or structural change of the model. A collective analysis of these change proposals is required to control the evolution of the process (CO), and it should first suggest a limited model evolution (only a single parameter of the model needs to be changed) or less marginal changes (when a combination of parameters needs to be changed). At this point, "New" after CO implies new method applications to each model variant and then new cycles (RA plus Com/RA). The conceptual validity of each modeling hypothesis and the logical validity of each new version of the formal model (using the terminology of Landry et al. [1983] and Oral and Kettani [1993]) are tested and evaluated at each iteration in relation to internal consistency and external validity criteria (Genard and Pirlot, 2002), and compared with other results of the original model and of the previously tested model variants (CO).

A certain sequence of improvements may not produce interesting results. If some validity criteria are not satisfied, CO can require a more complex change of some structural elements of the model, whereby one criterion is added at a time or modified, and if these changes are not sufficient, the logical structure of the model is analyzed and improved, or the problem formulation is changed and the model is connected to the new problem. Each modeling activity is completed with the application of an MC method to the new model version.

The passage from one model revision to another more viable one may be perceived as normal, logic, or essential, but a cycle of iterative tests is often required in this learning process. If the structure and parameters of a model evolve too quickly, they can be exposed to the risk of accepting badly formulated and not fully validated hypotheses or of losing the actual meaning of each modification because of, for example, the contemporary evolving of different elements.

Only when the modeling hypotheses and the last formal model have been considered acceptable ("Acc," after CO) can a robustness analysis be developed at a technical and collective level (ROB plus Com/ROB) in a new SISTI cycle, with the aim of testing the robustness of the collectively verified model (CO), rather than the robustness of the results, to stop the process or to activate a new SISTI cycle (Roy, 1998).

3.2.1 From Some Unusual Interventions to SISTI

SISTI should be applied in relation to problem situations in which a constructive approach is made impossible because a decision process is not yet activated and the MCDA intervention is required to facilitate its future activation. MC models, concepts, and procedures cannot be used as a communication tool for the participants in the decision process, but their potentiality in terms of transparency and possible inclusion of different knowledge elements can facilitate the structuring of a *conceptual model* (as it is defined in Landry et al. [1983]), which includes all the main aspects, requirements and uncertainties associated with the problem situation. The conceptual model structure is the same as that of an MC model (action and criteria), but the passage from a conceptual model to the development of a formal representation of the problem situation (*the formal model)*, specifically oriented toward the application of an MC method, is the result of cycles of method applications, result analysis, proposals of model improvement, and control of the process.

The documentation of the adopted logical paths and the main steps of the learning process facilitate a collective analysis of the process and above all of the evolution of the original problem formulation and importance of some aspects of the problem in the model. The often nonlinear path of a SISTI intervention stimulates new visions of the situation, orients the attention of stakeholders, or potentially involved actors, toward the specific nature of the decision problem, and sometimes anticipates some possible uses of data, indicators, and experts' judgments that can lead to the data and knowledge-acquisition process.

The idea that produced SISTI originated from the analysis of two old applications of MC methods in "unusual" MCDA interventions in research contexts, in relation to a pharmacological trial (Balestra et al., 2001) and to a heavy rain, flooding, and landslide event (Cavallo and Norese, 2001). When the clients are researchers involved in a new research topic, they may have data that need to be interpreted and translated into an organic and operational framework. The ELECTRE III method (Roy, 1978, 1990) was used in both of these interventions to structure and calibrate a "robust" MC model. The network of the involved actors included only domain experts and researchers, in both of the cases, and there was the possibility of using knowledge (from the clinical and biomedical research group in the first case and from the consequences of a specific event, in terms of erosion and slope instability, in the second) to create external criteria to evaluate the quality of the result. An application of ELECTRE III to the starting model produced completely unreliable results in both of the interventions. The analysis of these critical results was useful to recognize specific signs of weakness in the results of the ELECTRE III applications. At the same time, the external validity criteria confirmed the poor quality of the results. The process of arriving at a robust model was particularly long and difficult in the first case (Balestra et al., 2001). However, the acquired experience made the intervention described in Cavallo and Norese (2001) quicker and easier.

The factors that positively affected the two model-based interventions (above all the use of external criteria to evaluate the quality of the results and the identification of internal consistency/inconsistency signs) were proposed to students of degree and master courses and tested in laboratories (Norese, 2006).

At that point, the logic of the procedure that is here called SISTI was generalized and tested in a new intervention for a consultancy company that supplies different services to the Italian Aviation Authority and to several Italian airport concessionaire companies, to introduce an innovation in the airport services and to facilitate the decision process of the company's clients (Norese and Carbone, 2014). This time, the ELECTRE Tri method (Roy and Bouyssou, 1993) was used to generate a robust MC model that could be used by the consultancy company with its clients. Specific procedures of sensitivity analysis were introduced and tested.

Another SISTI application was developed in relation to the quality and shortcomings of the adopted resilience indices, and this application is presented in the next section.

3.3 MC MODELING OF RESILIENCE PROBLEMS

Over the last few decades, the concept of "resilience" has gained much ground in a wide variety of academic disciplines. The definitions are different and each of them includes different concepts, such as flexibility, adaptation, or reaction. Resilience would seem to be the answer to a wide range of problems and threats.

Several aspects of resilience were studied in the ANDROID—European Lifelong learning Programme to increase society's resilience to disasters of a human and natural origin . The definition of resilience that was proposed in the ANDROID Programme was rich and detailed: resilience is something that can grow in ourselves, in our family, and in our communities as the result of an educational activity addressed to the prevention and minimization of negative effects of adversities, natural events, disasters, and so on; the capacity of the administrators to face the risk of a catastrophe, their level of interest, resources, and efforts devoted to it (the social life sphere); the result of interactions among the environmental, sociopolitical, and economic factors that influence the various spheres of social life and activate the actors' awareness and involvement that are required to prevent and manage the effects of a disaster event.

An MC model was elaborated in 2014 in the ambit of the ANDROID Programme and in relation to a pilot case, to both underline the limits of the adopted resilience indices and to demonstrate, by means of the application of an MC method to a new resilience model, that MC models and methods "exist" and can be useful in resilience increasing processes (Scarelli and Benanchi, 2014). Starting from the results of this study, a SISTI application was carried out to test and orient the model in relation to a possible decision process of a territorial agency that needed structured knowledge before any resource allocation could be made.

A first SISTI iteration cycle (RA plus Com/RA plus CO and New) was activated to analyze the results of the ELECTRE III application to the model and to propose model improvements, new applications of the method, and new analyses of the results. When marginal changes to the model were considered insufficient to improve the model, a structural change of the model was implemented, and a second iteration cycle of SISTI was activated.

3.3.1 Original Model and First SISTI Iteration Cycle

A large number of indicators have been proposed in the literature and some resilience indices, which aggregate indicators, have been adopted. Scarelli and Benanchi (2014) proposed a different approach to the problem. As an alternative to the indices that combine different factors in a single synthetic value, they developed an MC model in which all the components were transparent, and they applied a multicriteria method, ELECTRE III, to synthesize the evaluations and rank the analyzed territorial units (21 municipalities of the Ombrone river basin in the Tuscany region in Italy; see Table 3.1 for the code used in the ELECTRE III applications and

TABLE 3.1 List of the Municipalities

Code	Municipality	Population	Code	Municipality	Population
RADI	Radicofani	1,148	MONTE	Monteroni d'Arbia	7,548
SART	Sarteano	4,679	RAPOL	Rapolano Terme	4,932
PIENZ	Pienza	2,231	CASTE	Castelnuovo Berardenga	8,081
SANQ	San Quirico d'Orcia	2,526	GAIOL	Gaiole in Chianti	2,333
CASTI	Castiglion d'Orcia	2,530	RADDA	Radda in chianti	1,715
MONTA	Montalcino	5,272	MONGG	Monteriggioni	9,165
MURLO	Murlo	2,116	SOVIC	Sovicille	8,882
BUONC	Buonconvento	3,197	RADIC	Radicondoli	1,019
SANGI	San Giovanni d'Asso	920	CHIUS	Chiusdino	1,944
TREQU	Trequanda	1,388	MONTI	Monticiano	1,412
ASCIA	Asciano	7,299			

the population of each municipality) from the most to the least resilient. The main reason for the choice of this MC method was its ability to pay particular attention to the uncertainty level that could have been associated to each indicator and the related evaluation.

The structure of an MC model (main aspects, or model dimensions, and criteria that analytically make each dimension operational) and the model parameters directly express the decision-makers' points of view. In a model oriented toward the use of an outranking method, the relative importance of the criteria (which only the decision makers can express to verify whether a concordance of "important" criteria exists and may facilitate a decision) and the veto thresholds, which model the need to control the risk of a high discordance between evaluations (complementary principle to the concordance principle in the ELECTRE methods and all the outranking methods), are the parameters that mainly require the expression of the decision-makers' points of view and preference systems. Other parameters (the indifference and preference thresholds) are often proposed to the decision makers as a technical proposal to reduce the uncertainty that may be associated with the data and the expressions of decision preference (see Roy, 1996).

The structure of the model in (Scarelli and Benanchi, 2014) consisted of two strategic aspects, or dimensions, and 14 criteria, the first 6 in relation to the environmental dimension, with almost the same importance as the other 8 criteria, in relation to the socioeconomic dimension (as indicated in the literature). The criteria were taken from literature, as was their relative importance (the "weights"), because of the absence of decision makers in this pilot case. The indifference and preference thresholds were instead elaborated in relation to the low quality of the data (taken from official databases but not so consistent with the nature of the required evaluations) that were used to evaluate the territorial units. No veto thresholds were introduced into the model.

A careful analysis of the ELECTRE III application results, and of their possible limits, was considered essential to verify whether this resilience model was sufficiently accurate, could give suitable explanations for the different situations in the Ombrone basin, and could be used to facilitate improvement actions (Norese et al., 2016; Norese and Scarelli, 2016).

The result of an ELECTRE III application is a classification of compared actions, from "best to worst," which is represented by a final partial graph (i.e., a pre-order that is developed as the intersection of the two complete pre-orders resulting from two distillation procedures, that is, the descendant procedure and the ascendant one [Figueira et al., 2005]). The final partial graph (see an example in Figure 3.2) can include different paths between the best and the worst actions, the longest of which can be visualized as the vertical and can be considered the main path, while each lateral path indicates a situation of incomparability for at least one couple of actions and underlines a distance (of one or more classes and sometimes even of several ones) between some action positions in the two distillations. The presence of different paths is more frequent when several actions are compared, and when the lateral paths appear in the intermediate part of the graph they do not compromise a clear definition of a ranking of the best actions (or of the worst, if the decision activity is oriented toward the elimination of the worst actions).

The number of lateral paths grows if the comparability of some actions is not so high, but a high number of lateral paths can sometimes be the sign of a difficult definition of some model parameters and above all of the veto thresholds. When MC modeling is particularly difficult (because the problem is new and complex or because modeling is requested in a laboratory for students who can be considered inexpert practitioners), the final partial graph often presents several incomparable actions that may be the consequence of incomplete or unstructured models or of nonconsistent or wrong definitions of some parameters (Norese, 2006).

In this case, the original model included a high (but not very high) number of actions (21 municipalities) and could have presented some elements of uncertainty because it was not created for a specific decision

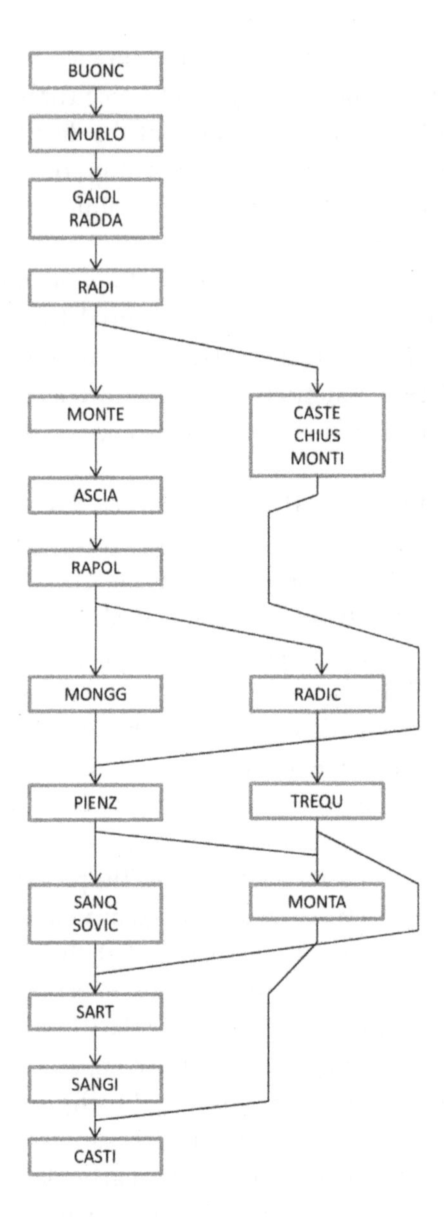

FIGURE 3.2 Final partial graph of the original model.

problem but only to improve future decision processes and because it synthesized logical inputs from literature and analytical inputs from the few available but not so reliable or consistent data.

The comparative visualization of some final partial graphs that result from ELECTRE III applications to model variants could be facilitated by indices that characterize the quality of the different results. Some indices were proposed in this SISTI application (always referring to the total number of actions):

- The number of actions in the longest path (VP), whose order is clear,
- The number of the actions in the lateral path (LP = 1 – VP), which can be considered an index of incomparability (presence of incomparable actions),
- An index of indifference and, therefore, of low/high discriminant power of the model, that represents the maximum number of actions that are assigned to the same class (II),
- Another index of incomparability (RI) that is expressed in relation to the action that is incomparable with the maximum number of other actions (RI is the number of actions that are incomparable with this action and can indicate a situation of substantial, also if local, incomparability).

The final partial graph in Figure 3.2, which resulted from the ELECTRE III application to the original model, was quite interesting, with a long main path (VP = 15/21), few actions in the same class (II = 3/21), a not so high presence of incomparable actions (LP = 6/21), only in the intermediate part of the graph, but a rather high index of incomparability (RI = 7/21) in relation to two municipalities (Radicondoli [RADIC] and Trequanda [TREQU)].

At that point, some small changes were introduced to improve certain indifference and preference thresholds that were too large, and the result changed considerably in relation to each small change. When some veto thresholds were introduced because the original model had not included any veto threshold, the final partial graph became disastrous (see Figure 3.3). The result of the ELECTRE III application to this model variant indicated the presence of several lateral paths and another critical and not so frequent element was present in the graph: just one best action was not included in this graph and a similar unusual situation emerged at the end of the ranking, with the consequence that the main path could not be identified.

All the parameter changes that were introduced, step by step, to improve the model produced different results, and when some changes

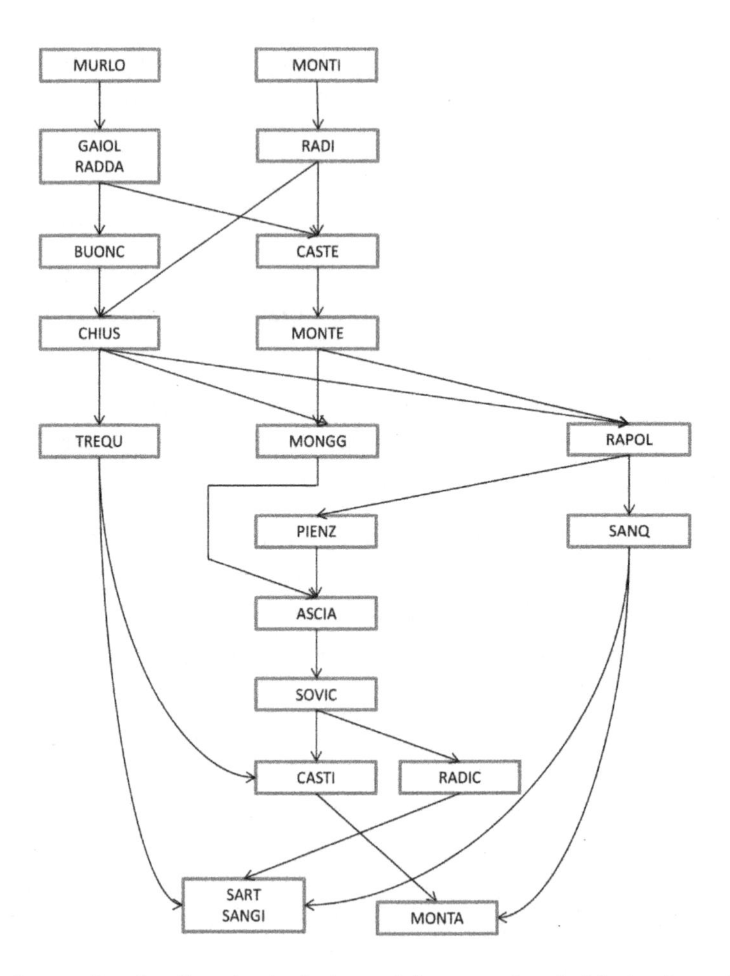

FIGURE 3.3 Result after the inclusion of the veto thresholds in the original model.

were introduced together it was evident how sensitive the result of the ELECTRE III application was to some parameter changes.

Each model variant was considered an improvement of some of the model parameters, but each variation produced less interesting results. At this point, the research team (which included the authors of the original model and the author of this paper together with a master thesis student) realized that no single marginal change could improve the result because the original choice of parameters had been conditioned by the difficult modeling context; the original model had been proposed only as a logical and analytical synthesis of the several inputs from the literature, without a specific decision problem having been defined, and there had been few and not so reliable or consistent data. A structural change to the model was considered the only possible course of action in SISTI and a new iteration cycle was activated.

3.3.2 The Second SISTI Iteration Cycle and a New Model

A new modeling logic was adopted to deal with a specific decision problem, in relation to the disaster resilience topic and to propose the results to policy makers and stakeholders involved in territorial processes. The original model, which was analyzed in terms of parameters (thresholds and modeling of the discordance principle) in the first iteration cycle, was then studied in terms of structure (the main conceptual aspects, or model dimensions, and a consistent family of criteria that analytically deal with these aspects [Roy, 1996]) and in relation to a possible evaluation process (choice of data-indicators-scales to be used in the evaluations).

A new and more specific decision problem was formulated by the research team, and three main aspects were identified to deal with the decision problem of a territorial agency that could allocate resources to improve the disaster resilience of the Ombrone river basin, in relation to the different *reaction capabilities* of the territorial units. In the Reaction capability model (see Table 3.2) some *Social aspects* and the *Ethical behavior* of the involved actors are included as main aspects because they can increase the reaction capability of each territorial unit; instead, *Risky behavior* can reduce the reaction capability. These three main aspects were considered the dimensions of the new model (set up to activate a new SISTI iteration cycle), which included 6 of the 14 criteria of the original model (Scarelli and Benanchi, 2014) and the related original evaluations.

The definition of the model parameters was facilitated by the knowledge that had been created in the previous parameter improvement iteration cycle (pertaining to the veto, indifference, and preference thresholds) and thanks to the clear structure of the new model that was able to facilitate a consistent definition of the criteria relative importance (in a logic of balancing the importance of the main aspects and of the criteria in relation to each aspect) and of the formulation of some policy scenarios. Three scenarios were introduced in relation to some possible policies: educating people to limit risky behavior, funding civil protection and training on how to react in the case of disaster activities, and funding landscape preservation and environmental protection activities. Each possible policy was associated with a different weight vector (the coefficients of the relative importance of the criteria) because of the absence of actual decision makers.

TABLE 3.2 Logical Structure of the Reaction Capability Model

Model Dimensions (Aspects and Possible Data)	**Criteria**
Risky behavior (anthropic impact on the environment, such as uncontrolled urbanization, cemented riverbanks, uncontrolled use of aquifer layers, high values of CO_2 emissions)	Uncontrolled *Urbanization*, which could limit rainfall absorption (rate of urbanized area, elaborated by means of GIS) *CO_2 emissions*, which could induce a high level of atmospheric contamination and alteration, as a sign of limited safety and risk awareness (source: Siena Province, Civil protection sector)
Social aspects (% of working women, scholastic attendance, population characteristics)	*Reaction time*, which is evaluated in terms of the ratio between the active population and the young plus old population (from the Demographic Dependency index, source: INSTAT, the National Institute of Statistics) *Progress in social life* (rate of unemployed women with respect to the total population, source: Siena Province, Report on the labour market)
Ethical behavior (Disaster prevention activities, such as naturalized river banks or education programs and awareness and interest in safeguarding the environment, such as differentiated waste, alternative energy use, high territorial desirability)	*Environmental awareness* (percentage of differentiated waste, source: Siena Province, Civil protection sector) *Safeguarding the considered area*, which could be expressed by the touristic attractiveness that motivates citizens and administrators to preserve the territorial qualities and to prevent any kind of negative impact (ratio between the touristic flows and resident population, source: Siena Province Tourist Office)

The ELECTRE III application to the new model (in relation to the balanced scenario) produced a somewhat interesting result (see the first graph in Figure 3.4). Both the first actions and the others at the end of the ranking again appeared clear: there were two municipalities (RADDA, MONGG) together in the first position and only one in the last (SART). There was a main path that contained 12 of the 21 municipalities, but some lateral paths (signs of incomparability situations) were present, above all in the middle part of the graph.

When the ELECTRE III application to the model was repeated with different importance of the criteria in relation to the three different policies/scenarios, the result changed and a comparative visualization of the

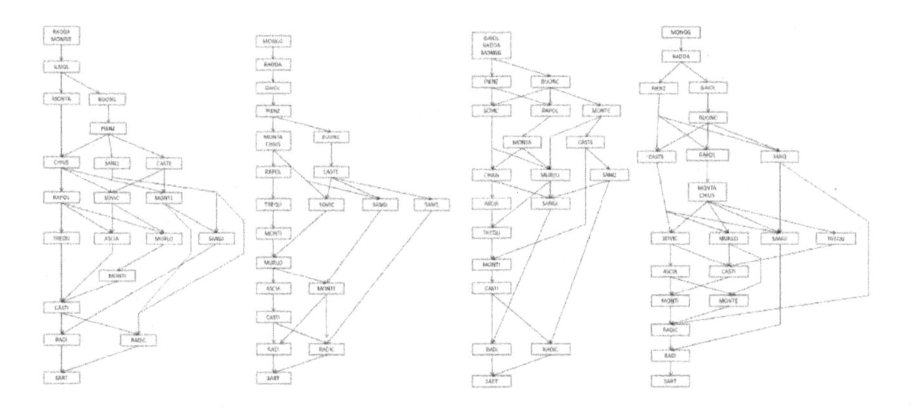

FIGURE 3.4 Final partial graphs of the new model in four policy scenarios.

results (see the other three graphs in Figure 3.4) facilitated the analysis of weakness and strong points of the model for future decisions in relation to unpredictable policies.

Five municipalities (RADDA, MONGG, GAIOL, BUONC, PIENZ) always remained in the first positions and four (SART, RADI, RADIC, CASTI) in the last positions in the ranking. The other 12 municipalities remained in the middle part, although there were some changes in their relative positions, and they always suffered from incomparability.

An analysis, which was conducted to test the sensitivity of the new model result (in relation to the balanced scenario), to veto changes (with the elimination of only one veto threshold each time) and to the introduction of a seventh criterion, produced the same result for the first and the last municipalities, and only small changes in the middle part.

The new model seemed quite interesting because the results were not so sensitive to the model parameters and allowed the study to came to some stable conclusions about the most and the least-resilient municipalities. However, although the application of ELECTRE III to the model was not able to produce a clear ranking of all of the 21 municipalities (the maximum number of actions in the main path is 14 in the four graphs in Figure 3.4), it confirmed their differentiation in three classes, as well as the sequence of the elements in the first and last classes. The several incomparability situations between the intermediate class elements limited the ranking capability and were unexplainable because they were not consistent with the high quality of the model parameters and the apparent comparability of the municipalities of the Ombrone basin.

At that point, an investigation was activated to obtain external criteria and to analyze new aspects pertaining to the intermediate class components and to the few of them that clearly changed their position in the analysis of the policies and scenarios. Most of the analyzed actions (see Table 3.1) are municipalities with just a few people, and the economic activities are predominantly of an agricultural and cattle breeding nature. Other municipalities are small, intact, and picturesque Middle-Age villages or small cities that are famous throughout the world for their wine or touristic attractiveness. The latter municipalities showed the most unexpected results (incomparable with a high number of other actions and important changes in their position in relation to different scenarios). A "natural" incomparability between these municipalities and several others in this river basin had to be accepted, and as a consequence, their ranking in the form of a complete order had to be considered almost impossible.

A different and more effective approach would have been to assign the units to different reaction capability and resilience categories, by means of a sorting method, for example ELECTRE Tri (Roy and Bouyssou, 1993) and only in a second step to generate rankings of the homogenous units of each category with different criteria in consideration of the nature of these units. Specific attention should be paid in a future model to some criteria that could have a specific meaning for some elements of the analyzed set and not for others, above all when the elements are very different. In most of the municipalities of Table 3.1 there is a very low level of industrialization (and low carbon dioxide [CO_2] emissions) and limited urbanization. Their very poor "waste differentiation" performance may be the consequence of the natural attitude of the inhabitants (whose economic activities are predominantly of an agricultural and cattle breeding nature) to reuse the waste that they produce on the farms to improve the fertility of the land or to heat their farms. Therefore, the municipalities may be characterized by good environmental awareness and territorial safeguarding, even though they do not differentiate their waste (the choice of a less ambiguous criterion is suggested in relation to this remark).

3.3.3 From This Result to Future Steps

This study started with an analysis of the results of an ELECTRE III application to a draft model and continued with an examination of the model elements that could negatively influence the result. Some hypotheses of parameter change were made and a sequence of ELECTRE III applications to the original model and the proposed variants was provisionally planned, because the analysis of each new result could orient the sequence of changes.

When each hypothesis of parameter change was analyzed and implemented, without any clear improvements in the results and with an evident worsening of the result at the end, possible limits in the model structure were analyzed and the structure was changed. A new model structure and a sequence of ELECTRE III applications to the new model, in relation to some policy/weight scenarios and parameter variants, produced more interesting results and the proposal of a new step of this SISTI application, which is currently underway (some first results are described in Norese, 2018).

Other SISTI applications have focused on the central role of modeling in decision aiding, in terms of both adopting a certain perspective in which uncertainty is accepted and flexibility is favored (as proposed in Vanderpooten, 2002), and the key role of the model in communication is underlined (see Landry et al., 1996; Genard and Pirlot, 2002).

The previously presented SISTI application was more oriented toward improving the practitioners' knowledge of how a result should be interpreted, critically analyzed, and used to improve a model. Modeling is a very hard task, but other tasks are not easy either. Sensitivity analysis is an important activity but a clear and detailed procedure does not exist in literature (at least in the MC field), and formulas "similar" to those proposed in the sensitivity analysis of an optimal solution are not present in the multicriteria analysis field.

The incomparability relation is a basic component of the outranking relation, but the interesting visualization of incomparability in a partial final graph is often too difficult to be understood, at least for inexpert practitioners and young researchers. For this reason, some description of the graphs' meaning and a proposal of indices of result quality were proposed in this presentation of a SISTI application to improve and generalize visualization and critical analysis of this kind of result. An Software (SW) tool could include and visualize parameters that describe the main elements of a final partial graph and its evolution during the modeling process and could also propose other visualization tools, such as the Surmesure diagram that is described in Rogers et al. (2000). Some difficulties that the end users encounter when they use ELECTRE III are linked to the not so easy interpretation of its results, above all when the problem is new and several actions are compared.

The SISTI application to the resilience problems is not concluded, but the last step, which is currently underway, has the aim to arrive at a complete and well-documented application that can be proposed as a way to improve model and problem formulation by means of an informed use of MC methods and as a stimulus to improve the effectiveness of some MC SW tools in the modeling and result analysis activities.

3.4 CONCLUSIONS

Decision aiding and interaction with the participants in a decision process become difficult when a problem situations is so new that a decision process is not yet activated and a formal decision system does not exist, but there is already the perception of an unstructured decision problem.

SISTI is a methodological approach that can be used to deal with these situations by means of cyclic applications of an MC method to a model, which has been developed without decision makers, and analyses of the results of these applications, to study how the uncertainty of an output can be connected to different sources of uncertainty in the model structure and parameters, which are the inputs of the method application. Some alternative assumptions on the choice of the parameters (or on the model structure), which could generate uncertainty or criticalities in the results, can be tested to determine their impact on the results and then used. An identification of the inputs that cause significant uncertainty in the output should orient attention toward improving a model that cannot be validated by means of a natural interaction between decision makers and analysts. A comparison of the results, and above all, of their unusual elements, with a different type of information about these elements (the external criteria) is essential to orient each cycle of modeling.

SISTI is a decision aid approach that *poses rather solves the problem* (Roy, 1996) during a study that simulates an interaction with decision makers and is instead developed together with a few actors who perceive the need to understand and propose structured elements for later phases of a still-not activated decision process. SISTI is not a new MC method; it is instead a proposal of using an MC method with the same aim of a problem structuring method, to structure a new and complex problem and to elaborate and validate an open model (Vanderpooten, 2002) when decision makers do not exist, cannot participate in, or do not want to be involved in the decision-aid process. The study is not directly oriented toward a forthcoming and hard decision, but it has the aim of stimulating new points of view and perceptions of the problem situation and of increasing the actors' knowledge of the situation by means of the understanding of the expected and unexpected relationships between a model and the always clearer results of an MC method application.

The main drawback of SISTI is related to time: a reliable prevision of how many iterations and how much time is required to arrive at a good model, which would be able to express useful knowledge of the problem, is difficult or even impossible. An essential prerequisite is the presence of at least some knowledge, which should be used to test the model that evolves in SISTI, or the existence of a great deal of data, from which to extract knowledge and use it for the problem analysis.

Another shortcoming is that this approach, which has been applied several times to clarify and structure complex situations, is not general enough to be proposed to young researchers and inexpert practitioners, to help them understand what a "good" model is and how the robustness of their conclusions can be improved. The students' reactions to this message underlined this problem in the past and the need for an SW tool that could play an important role in this modeling process, by facilitating the comparison of the different results and the visualization of the impact on the results that is generated by each parameter assumption or modeling scenario.

The integration of some "modeling assistants" that can be essential for MC decision aid to improve the important role that an SW tool could play (but currently does not play) in the modeling process was proposed during the 15th Decision Deck Workshop in Lisbon (Norese et al., 2018). This possible improvement, together with the description of a complete and not so complex case as a guideline for practitioners and young researchers, should make SISTI more and more general and easier to apply.

REFERENCES

Balestra G., Norese M.F., Knaflitz M. (2001) Model structuring to assess the progression of muscular dystrophy, In A. Colorni, M. Parruccini, B. Roy (Eds.), *A-MCD-A – Aide Multicritère à la Décision (Multiple Criteria Decision Aiding)*. European Commission Joint Research Centre, EUR Report, Luxembourg, 31–46.

Bana e Costa C.A., Ensslin L., Correa E.C., Vansnick J.-C. (1999) Decision support systems in action: Integrated application in a multicriteria decision aid process. *European Journal of Operational Research*, 113, 315–335.

Belton V., Ackermann F., Shepherd I. (1997) Integrated support from problem structuring through to alternative evaluation using COPE and VISA. *Journal of Multi-Criteria Decision Analysis*, 6, 115–130.

Belton V., Stewart T.J. (2010) Problem Structuring and Multiple Criteria Decision Analysis. In M. Ehrgott, J.R. Figueira, S. Greco (Eds.) *Trends in Multiple Criteria Decision Analysis*, International Series in Operations Research and Management Science 142, Springer-Verlag, Heidelberg, Germany, 209-239.

Cavallo A., Norese M.F. (2001) GIS and multicriteria analysis to evaluate and map erosion and landslide hazard. *Informatica*, 12(1), 25–44.

de Montgolfier J., Bertier P. (1978) *Approche Multicritère des problèmes de décision*, Editions. Hommes et Techniques, Paris.

Figueira J., Mousseau V., Roy B. (2005) ELECTRE methods, In S. Greco (Ed.), *Multiple Criteria Decision Analysis: State of the Art Surveys.* Springer, Heidelberg, Germany, 133–153.

Genard J.-L., Pirlot M. (2002) Multi-criteria decision-aid in a philosophical perspective. In D. Boyssou, E. Jacquet-Lagrèze, P. Perny, R. Slowinski, D. Vanderpooten, P. Vincke (Eds.), *Aiding Decisions with Multiple Criteria: Essays in Honour of Bernard Roy.* Kluwer Academic, Dordrecht, the Netherlands, 89–117.

Landry M., Banville C., Oral M. (1996) Model legitimization in operational research: Model validation in operations research. *European Journal of Operational Research*, 14, 207–220.

Landry M., Malouin J.L., Oral M. (1983) Model validation in operations research. *European Journal of Operational Research*, 92, 443–457.

Montibeller G., Belton V., Ackermann F., Ensslin L. (2008) Reasoning maps for decision aid: An integrated approach for problem-structuring and multicriteria evaluation. *Journal of the Operational Research Society*, 59, 575–589.

Norese M.F. (2006) Multicriteria modeling and result analysis. *Newsletter of the European Working Group "Multiple Criteria Decision Aiding,"* 3(14), 10–14.

Norese M.F. (2018) How ELECTRE Tri and the combined action of two SW tools can be used to create a robust model. *Newsletter of the European Working Group "Multiple Criteria Decision Aiding,"* Series 3, 38,4–11.

Norese M.F., Montagna F., Riva S. (2008) A multicriteria approach to support the design of complex systems. *Foundations of Computing and Decision Sciences*, 33(1), 53–70.

Norese M.F., Carbone V. (2014) An application of ELECTRE Tri to support innovation. *Journal of Multi Criteria Decision Analysis*, 21, 77–93.

Norese M.F., Mustafa A., Scarelli A. (2016) New frontiers for MCDA: From several indicators to structured models and decision aid processes. *Newsletter of the European Working Group "Multiple Criteria Decision Aiding,"* 3 (34), 1–8.

Norese M.F., Scarelli A. (2016) Decision aiding in public policy generation and implementation: A multicriteria approach to evaluate territorial resilience. *Territorio Italia*, 2, 71–90.

Norese M.F., Profili A., Scarelli A. (2018) How modelling assistants can be integrated in the SW tools to improve the effectiveness of decision aiding. 15th Decision Deck Workshop, September 26, 2018, Lisbon, https://www.decision-deck.org/news/newsletter1.html.

Oral M., Kettani O. (1993) The facets of the modeling and validation process in operations research. *European Journal of Operational Research*, 66, 216–234.
Petkov D., Petkova O., Andrew T., Nepal T. (2007) Mixing multiple criteria decision making with soft systems thinking techniques for decision support in complex situations. *Decision Support Systems*, 43, 1615–1629.
Rosenhead J. (1996) What's the problem? An introduction to problem structuring methods. *Interfaces*, 26(6), 117–131.
Rosenhead J., Mingers J. (Eds) (2001) *Rational Analysis for a Problematic World Revisited: Problem Structuring Methods for Complexity, Uncertainty and Conflict*. Wiley, Chichester, UK.
Roy B. (1978) Electre III: Un algorithme de classements fondé sur une représentation floue en présence de critères multiples. *Cahier du CERO*, 20, 24–27.
Roy B. (1981) The optimization problem formulation: Criticism and overstepping. *Journal of the Operational Research Society*, 32(6), 427–436.
Roy B. (1985) *Méthodologie Multicritère d'Aide à la Décision: Méthodes et Cas*. Economica, Paris, France.
Roy B. (1990) The outranking approach and the foundations of ELECTRE methods. In C.A. Bana e Costa (Ed.) *Readings in Multiple Criteria Decision Aid*. Springer-Verlag, Heidelberg, Germany, 155–184.
Roy B. (1996) *Multicriteria Methodology for Decision Aiding*. Kluwer Academic, Dordrecht, the Netherlands.
Roy B. (1998) A missing link in OR/DA: Robustness analysis. *Foundations of Computing and Decision Sciences*, 23(3), 141–160.
Roy B., Bouyssou D. (1993) *Aide multicritère à la décision: mèthodes et cas*. Economica, Collection Gestion, Paris, France.
Rogers M., Bruen M., Maystre L. (2000) *ELECTRE and Decision Support: Methods and Applications in Engineering and Infrastructure Investment*. Kluwer Academic Publishers, Boston, MA.
Scarelli A., Benanchi M. (2014) Measuring resilience on communalities involved in flooding Ombrone river. *Procedia Economics and Finance*, 18, 948–958.
Simon H.A. (1960) *The New Science of Management Decision*. Harper & Brothers, New York, NY.
Simon H.A. (1991) Bounded rationality and organisational learning. *Organisation Science*, 2, 125–139.
Stewart T.J., Joubert A., Janssen R. (2010) MCDA framework for fishing rights allocation in South Africa. *Group Decision and Negotiation*, 19, 247–265.
Tsoukias A. (2007) On the concept of decision aiding process: An operational perspective. *Annals of Operations Research*, 154, 3–27.
Vanderpooten D. (2002) Modelling in decision aiding. In D. Boyssou, E. Jacquet-Lagrèze, P. Perny, R. Slowinski, D. Vanderpooten, P. Vincke (Eds.), *Aiding Decisions with Multiple Criteria: Essays in Honor of Bernard Roy*. Kluwer Academic, Dordrecht, the Netherlands, 195–210.

CHAPTER 4

Applying Intangible Criteria in Multiple-Criteria Optimization Problems

Challenges and Solutions

Marina V. Polyashuk

CONTENTS

4.1 INTRODUCTION

The classical multiple-criteria optimization model under certainty consists of three major components: decision space, criteria space, and criteria mapping. Decision space is usually viewed as a multidimensional space in which all possible alternatives (decisions) are defined as vectors of parameter values. For example, in the investment portfolio selection problem, the decision space can be considered as a set of all possible portfolios, where a vector of investments represents a potential decision (Polyashuk, 2005). Criteria space, on the other hand, contains possible vectors of criteria values, and this is the space where the decision-maker's preferences are defined. In the example of investment portfolio selection, the simplest criteria space (based on Modern Portfolio Theory) would consist of possible two-dimensional vectors, where a vector shows the expected return and risk in the form of variance of the return generated by a specific portfolio of assets (Markowitz, 1959). The third component of the classical multiple-criteria model, criteria mapping, assigns to every feasible decision in the decision space the corresponding vector of criteria evaluations, so that the decision maker can use them to find preferred solution(s). Creating a multiple criteria optimization model is a complicated and challenging process (see, for example, Zeleny, 2011). For the model to work, we need a clear understanding of each of its three components, and a major challenge comes from determining criteria and their values. This is the area that produces the most arguments and disagreements among both researchers and practitioners in multiple-criteria optimization, especially when it is extremely difficult to combine criteria that are entirely different in their nature.

Among various considerations, if some of the criteria cannot be measured in an objective fashion, then the formulation of an optimization model becomes a tremendous challenge, and more importantly, decision analysts are compelled to create biased, unproven models to describe the criteria set. In a typical approach to evaluating intangibles, we often try to put quality in a numerical form. How often do we participate in "evaluating" unmeasurable things on the scale from 1 to 5? Or from 0 to 10? Each evaluation rubric and each scale may yield a different optimal solution, even if we agree on the optimization model. Constructing a "qualitative value scale" was addressed in Belton and Stewart (2002) and is a thoughtful process of elicitation of the decision-maker's evaluations; however, this methodology works only under strict assumptions regarding the decision maker and may lead to inconsistencies.

In the current work, we first discuss major challenges in forming the criteria set and supplying each criterion with a mechanism for evaluating alternative decisions against it. Further, we consider ways of incorporating qualitative, intangible objectives into a process of decision making with multiple criteria. Our approach requires a priori understanding of the role that these objectives play in determining decision-maker's preferences, and this approach leads to a way of considering qualitative aspects of alternatives without forcing either side (the decision maker and the decision analyst) to create a biased model going beyond actual approximation of preferences. Finally, we will show that a multiple-criteria model with more than one criteria set may resolve difficult situations with intangible criteria.

4.2 METHODOLOGICAL AND PHILOSOPHICAL CHALLENGES IN APPLYING INTANGIBLE OBJECTIVES

Difficulties in applying various objectives may stem from situations in which definitions are ambiguous and allow for different interpretations. Let us demonstrate this by the examples related to the problem of choosing an optimal investment portfolio. Consider two criteria used in evaluating investment portfolios: (1) the expected return and (2) the level of diversification. Analysts and decision makers may disagree on probability distributions; nevertheless, once certain assumptions are agreed on, it becomes possible to measure the expected return. On the other hand, the level of diversification represents a different challenge because there is no exact, universally accepted way to define it as a measurable characteristic. Examples of expected return and diversification demonstrate the distinction between tangible and intangible, measurable and not measurable.

An analogy of the relationship between qualitative and quantitative characteristics can be found in probability theory when considering subjective probability compared with theoretical or empirical probability. Subjective probability is often based on comparing likelihoods of different events, whereas both theoretical and empirical probabilities are based on measurements: theoretical, on computing the probability of an event as the sum of probabilities of its outcomes, and empirical, on relative frequency of the observed event. See, for example, the distinction between qualitative and quantitative probabilities (Savage, 1972). Because the qualitative probability is established as a relation (such as one outcome is less likely than the other), then the common additive probability model does not apply, just as ordinal scale cannot be naturally converted into an interval or ratio scale. Similarly, converting qualitative evaluations into a numerical form, especially for performing operations with them such as finding sums or differences, does not have a solid foundation. Consider, for example, students' evaluations of teacher performance where there are five possible answers to each question: strongly disagree, disagree, neutral, agree, and strongly agree. These qualitative values are then assigned numbers 1 through 5; these numbers may be averaged for one teacher across the questions and for all teachers across the unit, and they may be used to compare teachers' performance or to evaluate progress in a single teacher's professional ability. All of this is done while there is no meaningful numeric representation of the difference between agreeing and disagreeing, being neutral or strongly agreeing. These differences and the assigned numerical values have no meaning except being positive, negative, or zero. On the other hand, the same surveys could be used in a meaningful way by making a conclusion that the students have a good opinion of their teacher if a majority of them agree or strongly agree with the important aspects of the teacher's performance. There are legitimate ways of measuring teacher's performance, such as finding the difference in students' scores on standardized tests at the beginning and at the end of a school year. These examples, just like the examples related to investment portfolios, demonstrate a distinction between qualitative, intangible values and quantitative, tangible ones.

Although various criteria might have a different nature, in practical applications we often see attempts of combining (i.e., aggregating) them, regardless of being quantitative or qualitative, in some sort of numeric form quantifying decision-maker's preferences. This represents another challenge in applying multiple criteria: there is no efficient way of combining quantitative, tangible and qualitative, intangible criteria in most widely used multicriteria models such as a value function, reference point method, and so on, unless one tries to squeeze qualitative criteria into some quantified version. Linear models are most common in multicriteria mathematical programming (Doumpos et al., 2001), but these models also require numerical representation of all decision-maker's objectives. When one combines various types of criteria in a single additive model, it tends to prevent this model from being an accurate reflection of reality and might cause failure in finding truly optimal solutions. Furthermore, attempts of creating more tractable optimization models such as models with weighted criteria become more open to criticism and questions about the character and adequacy of such models. Eliciting weights based on relative importance of criteria is difficult even for measurable criteria (see Belton and Stewart, 2002; Alfares and Duffuaa, 2009) and becomes even more problematic when part of the criteria set consists of qualitative characteristics of decisions.

So far, we have been discussing methodological challenges that arise when qualitative, intangible criteria are part of the consideration in the decision-making process. Philosophical challenges are also many, starting with the fact that the same problem, solved with different approaches, will lead to different results, which all will be called "optimal" by their proponents. Which one of these proposed solutions is genuinely optimal and reflects the decision-maker's preferences? This thought brings us back to the foundations of decision theory (Arrow, 1959; Sen, 1971; Plott, 1976), choice functions and their properties, and correspondence between those properties and the existence of structures that would be consistent with actual preferences of the decision maker such as a value function or a binary relation. Are we helping decision makers to make up their minds, or are we doing it for them when applying various models and methods? After all, human beings are making complicated decisions every day without consulting with analysts, such as choosing a cup of coffee at Starbucks, selecting an outfit, picking exercises for the next lesson, choosing a contractor, buying a car or a home, and so on. Maybe multicriteria models, especially with some qualitative criteria, should be just founded on several basic principles while narrowing the set of choices for the decision maker to continue with the process? In the current work, we are trying to get closer to answering these questions by proposing several possible ways to handle a combination of quantitative and qualitative, tangible and intangible objectives during multi-criteria decision making.

4.3 THEORETICAL FOUNDATION FOR TREATMENT OF QUALITATIVE CRITERIA

The challenges that arise from some criteria being qualitative suggest that, in practical applications, we should treat these criteria differently, and therefore, we should distinguish different groups among multiple criteria. Before we describe formal procedures for treating qualitative criteria, let us review similar concepts of criteria treatment.

Lexicographically ordered groups of criteria (Kostreva and Polyashuk, 1993) were introduced through a type of binary relation, which induces the choice of optimal solutions as a succession of choices of Pareto-optimal solutions with respect to these groups, in the order of their importance. Partition of the criteria set S into two subsets, S_1 and S_2, where the criteria set S_1 is more important than the criteria set S_2 (Polyashuk, 2005) allowed developing a portfolio selection procedure with this type of criteria set. Seemingly unrelated to relative criteria importance, the concept of preferential independence (Keeney and Raiffa, 1976) represents situations when decision-maker's preferences with respect to one set of criteria (attributes) are not affected by the values with respect to the complementary set of criteria (attributes). These concepts will be stepping-stones in formalizing our approach to creating multi-criteria models with qualitative criteria. To continue, we will need to introduce several definitions.

Definition 4.3.1

Let $\mathbb{R}^m$ be the decision space: the set containing alternative solutions as vectors of m parameter values. Let $\widetilde{A}$ be the set of feasible solutions in the decision space: $\widetilde{A} \subseteq \mathbb{R}^m$.

Definition 4.3.2

Let $\mathbb{R}^n$ be the criteria space with n criteria: the set containing vectors of ordered evaluations of alternative solutions with respect to n criteria $K_1, K_2, \ldots K_n$. Let $\mathbf{F}: \widetilde{A} \rightarrow \mathbb{R}^n$ be the criteria mapping, which assigns to each element $\mathbf{a} \in \widetilde{A}$ a vector of n criteria values $\mathbf{F}(\mathbf{a}) \in \mathbb{R}^n$. Let: $A = \mathbf{F}(\widetilde{A})$ be the image of $\widetilde{A}$ under $\mathbf{F}$.

Definition 4.3.3

Let there exist the choice of preferred vectors $C(X) \subseteq X$ for any nonempty $X \subseteq \mathbb{R}^n$ such that $C(X) \neq \varnothing$ – a choice function $C(\cdot)$ defined on the power set $\mathbf{P}(A)$. An optimal solution chosen from a subset of feasible solutions $Y \subseteq \widetilde{A}$ is defined as any element $\mathbf{y} \in Y$ such as $\mathbf{y} \in \mathbf{F}^{-1}(C(\mathbf{F}(Y)))$.

Definition 4.3.4

Pareto relation $\prec \subseteq \mathbb{R}^n \times \mathbb{R}^n$ is defined as follows: $\forall \mathbf{x}, \mathbf{y} \in \mathbb{R}^n$, $\mathbf{x} \prec \mathbf{y}$ if and only if $x_i \leq y_i, \forall i = 1, 2, \ldots, n$ and $\exists j \in \{1, 2, \ldots, n\}$ such that $x_j < y_j$, where $\mathbf{x} = (x_1, x_2, \ldots x_n)$ and $\mathbf{y} = (y_1, y_2, \ldots y_n)$. The meaning of $\mathbf{x} \prec \mathbf{y}$ is that $\mathbf{x}$ is less preferred than $\mathbf{y}$.

Definition 4.3.5

Sen (1971). Preference relation induced by the choice function $C(\cdot)$ is relation $R_C \subseteq \mathbb{R}^n \times \mathbb{R}^n$ defined as follows: $\forall \mathbf{x}, \mathbf{y} \in \mathbb{R}^n$, $\mathbf{x} R_C \mathbf{y}$ ($\mathbf{x}$ is less preferred than $\mathbf{y}$) if and only if $\forall X \subseteq \mathbb{R}^n$ such that $\mathbf{x}, \mathbf{y} \in X$, $\mathbf{x} \notin C(X)$, and $\mathbf{y} \in C(X)$.

Definition 4.3.6

Kostreva and Polyashuk (1993). Relation $R^L \subseteq \mathbb{R}^n \times \mathbb{R}^n$ is the relation based on lexicographic-ordered groups of criteria $\{T_1, T_2, \ldots T_l\}$ where $\{T_1, T_2, \ldots T_l\}$ is a partition of the set of criteria K in which each group T_j is more important than the group T_{j+1}, if and only if $\forall \mathbf{x}, \mathbf{y} \in \mathbb{R}^n$ $\mathbf{x} R^L \mathbf{y} \Leftrightarrow \exists j \in \{1, 2, \ldots, l\}$ such that $\mathbf{x} \prec_{T_j} y$, and $\forall i < j$, $\mathbf{x}$ is not comparable to $\mathbf{y}$ with respect to relation $\prec_{T_i}$.

Definition 4.3.7

Let $\mathbf{x} \in \mathbb{R}^n$ where $\mathbf{x} = (x_1, x_2, \ldots x_n)$; let $K = \{K_1, K_2, \ldots K_n\}$ and $T \subseteq K$: $T = \{K_{i_1}, K_{i_2}, \ldots K_{i_r}\}$. The restriction of $\mathbf{x} \in \mathbb{R}^n$ to T is $\mathbf{x}|_T = (x_{i_1}, x_{i_2}, \ldots x_{i_r})$. An extension of the vector $\mathbf{a} = (x_{i_1}, x_{i_2}, \ldots x_{i_r})$ to K, $\mathbf{a}|^K$, is a vector $\mathbf{y} = (y_1, y_2, \ldots y_n) \in \mathbb{R}^n$ such that $a_{i_p} = y_{i_p}$, $\forall p = 1, 2, \ldots r$.

Definition 4.3.8

Let $K = \{K_1, K_2, \ldots K_n\}$ and $T \subseteq K$. For any set $A \subseteq \mathbb{R}^n$, the restriction of A to T is $\left\{\mathbf{x}\middle|_T = (x_{i_1}, x_{i_2}, \ldots x_{i_r}) \middle| \mathbf{x} \in A\right\}$. The extension of a set $Y \subseteq \mathbb{R}^r$ to K is the set $\left\{\mathbf{y} = \mathbf{a}|^K \middle| \mathbf{a} \in Y\right\}$.

Definition 4.3.9

Polyashuk (2005). Suppose that for a given criteria set $T \subseteq K$ there exists a choice function $C|_T(\cdot)$ on the power set $\mathrm{P}(A|_T)$ where $A|_T = \{\mathbf{x}|_T | \mathbf{x} \in A\}$. Then the set of criteria T is more important for the decision maker than its complement $K - T$ if for any $X \subseteq \mathbb{R}^n$, $C(X) \subseteq C\left((C_T(X|_T))|^K\right)$.

Definition 4.3.10

Keeney and Raiffa (1976). Given a preference relation R, the set of criteria ("attributes") T where $T \subseteq K$, is preferentially independent of its complementary set of criteria ("attributes") $K - T$ if and only if $\forall \mathbf{x}, \mathbf{y}, \mathbf{u}, \mathbf{w} \in \mathbb{R}^n$ such that $\mathbf{x}|_T = \mathbf{u}|_T$, $\mathbf{y}|_T = \mathbf{w}|_T$, $\mathbf{x}|_{K-T} = \mathbf{y}|_{K-T}$, and $\mathbf{u}|_{K-T} = \mathbf{w}|_{K-T}$, $\mathbf{x} R \mathbf{y} \Rightarrow \mathbf{u} R \mathbf{w}$.

Preferential independence of T of its complementary set is interpreted as follows: although the criteria values with respect to the complementary set are fixed, decision-maker's preferences do not depend on these common fixed values but only on the values with respect to criteria set T.

Definition 4.3.11

Given a preference relation R, the set of criteria set T where $T \subseteq K$, is **strictly** preferentially independent (or strictly independent) of $K-T$ if and only if $\forall \mathbf{x}, \mathbf{y}, \mathbf{u}, \mathbf{w} \in \mathbb{R}^n$ such that $\mathbf{x}|_T = \mathbf{u}|_T$ and $\mathbf{y}|_T = \mathbf{w}|_T$, $\mathbf{x}R\mathbf{y} \Rightarrow \mathbf{u}R\mathbf{w}$.

We may interpret Definition 4.3.12 in this way: S is strictly independent of T suggests that if the decision maker prefers $\mathbf{y}|_T$ to $\mathbf{x}|_T$, then even if we arbitrarily change the values with respect to set $K-T$, the decision maker will still prefer $\mathbf{y}$ to $\mathbf{x}$. It is clear from the definitions that if the set of criteria set T is **strictly** independent of $K-T$, then T is **preferentially** independent of $K-T$. However, strict independence suggests precedence of criteria set T over $K-T$, which establishes a strict ordering on the criteria set K. Let us consider the following example to clarify the difference between the definitions.

Example

Let $K = \{K_1, K_2, K_3, K_4\}$, $T = \{K_1, K_2\}$, and $K-T = \{K_3, K_4\}$. Suppose $\mathbf{x} = (1,2,5,8)$, $\mathbf{y} = (3,4,5,8)$, and $\mathbf{x}R\mathbf{y}$. Then, if T is preferentially independent of $K-T$ (Definition 4.3.10), it follows that $\mathbf{u}R\mathbf{w}$ where $\mathbf{u} = (1,2,3,0)$ and $\mathbf{w} = (3,4,3,0)$. If T is strictly independent of $K-T$ (Definition 4.3.11), then $\mathbf{x}R\mathbf{y}$ implies that $\mathbf{u}R\mathbf{w}$ also for $\mathbf{u} = (1,2,10,5)$ and $\mathbf{w} = (3,4,1,1)$.

Theorem 4.1

In the case of Pareto relation $\prec \subseteq \mathbb{R}^n \times \mathbb{R}^n$, each nonempty subset T of the criteria set K is preferentially independent (in the sense of Definition 4.3.10) of its complement $K-T$.

Proof

Let $T \subseteq K$, $T \neq \emptyset$, where $T = \{K_{i_1}, K_{i_2}, \ldots K_{i_r}\}$. Without loss of generality, let us assume that $\{i_1, i_2, \ldots i_n\} = \{1,2,\ldots,n\}$. Then $T = \{K_1, K_2, \ldots, K_r\}$ and $K-T = \{K_{r+1}, K_{r+2}, \ldots, K_n\}$. Suppose $\forall \mathbf{x}, \mathbf{y}, \mathbf{u}, \mathbf{w} \in \mathbb{R}^n$ and $\mathbf{x}|_T = \mathbf{u}|_T$, $\mathbf{y}|_T = \mathbf{w}|_T$, $\mathbf{x}|_{K-T} = \mathbf{y}|_{K-T}$, and $\mathbf{u}|_{K-T} = \mathbf{w}|_{K-T}$. Let $\mathbf{x} \prec \mathbf{y}$. Then $x_i \leq y_i, \forall i = 1,2,\ldots,n$ and $\exists j \in \{1,2,\ldots,n\}$ such that $x_j < y_j$. Since $\mathbf{x}|_{K-T} = \mathbf{y}|_{K-T}$, $x_i \leq y_i, \forall i \in \{1,2,\ldots,r\}$ and $\exists j \in \{1,2,\ldots,r\}$ such that $x_j < y_j$. Since $\mathbf{x}|_T = \mathbf{u}|_T$ and $\mathbf{y}|_T = \mathbf{w}|_T$, we have $u_i \leq w_i, \forall i \in \{1,2,\ldots,r\}$ and $\exists j \in \{1,2,\ldots,r\}$ such that $u_j \leq w_j$; together with the fact that $\mathbf{u}|_{K-T} = \mathbf{w}|_{K-T}$, it implies that $\mathbf{u} \prec \mathbf{w}$. Therefore, $\mathbf{x} \prec \mathbf{y} \Rightarrow \mathbf{u} \prec \mathbf{w}$.

Hence, given a nonempty subset T of the criteria set K and Pareto relation $\prec$, T is preferentially independent of its complement $K-T$. *Q.E.D.*

Theorem 4.2

If a subset of criteria set T where $T \subseteq K$, is more important for the decision maker than its complement $K-T$ in the sense of Definition 4.3.9, then the group of criteria T is strictly independent of the complementary group $K-T$.

Proof

Suppose $T \subseteq K$ and $T \neq \emptyset$; suppose that T is more important than its complement $K-T$ in the sense of Definition 4.3.9: for the choice function $C(\cdot)$, there exists a choice function $C_T(\cdot)$ such that $\forall X \subseteq \mathbb{R}^n$, $C(X) \subseteq C\left((C_T(X|_T))|^K\right)$. Let R_C be the preference relation induced by the given choice function $C(\cdot)$.

Assume that $\mathbf{x}, \mathbf{y}, \mathbf{u}, \mathbf{w} \in \mathbb{R}^n$ with $\mathbf{x}|_T = \mathbf{u}|_T$ and $\mathbf{y}|_T = \mathbf{w}|_T$, and that $\mathbf{x} R_C \mathbf{y}$. Then $\exists X \subseteq \mathbb{R}^n$ such that $\mathbf{x}, \mathbf{y} \in X$, $\mathbf{x} \notin C(X)$, and $\mathbf{y} \in C(X)$. Then $\mathbf{x} \notin C_T(X)$ and $\mathbf{y} \in C_T(X)$. Since $\mathbf{x}|_T = \mathbf{u}|_T$ and $\mathbf{y}|_T = \mathbf{w}|_T$, this implies that $\mathbf{u} \notin C_T(X)$ and $\mathbf{w} \in C_T(X)$, so $\mathbf{u} \notin C(X)$ and $\mathbf{w} \in C(X)$. Therefore, $\mathbf{u} R_C \mathbf{w}$. Hence, given a nonempty subset T of the criteria set K, which is more important than its complement $K-T$, in the sense of Definition 4.3.9, T is preferentially independent of its complement $K-T$. *Q.E.D.*

Corollary 4.3

If the decision maker's preferences are consistent with relation R^L based on lexicographic-ordered groups of criteria $\{T_1, T_2, \ldots T_l\}$, then $\forall p \in \{1, 2, \ldots, l-1\}$, group of criteria $\bigcup_{i=1}^{p} T_i$ is strictly independent of the group $\bigcup_{i=p+1}^{n} T_i$.

The proof follows from Theorem 4.2 and the fact that the set of nondominated points with respect to relation R^L is equal to $Max_{\prec_{Tl}}(Max_{\prec_{Tl-1}}(\ldots(Max_{\prec_{T1}}))\ldots)$ where $Max_{\prec_{Tj}}$ denotes choice of Pareto-optimal vectors with respect to T_j (see Kostreva and Polyashuk, 1993).

Theorem 4.4

Suppose that strict independence relation $I \subseteq \mathrm{P}(K) \times \mathrm{P}(K)$ is defined on the power set $\mathrm{P}(K)$ where K is the criteria set and that the decision-maker's preferences are described by a (nonempty) strict partial ordering. Then relation I is asymmetric.

Proof

Let R, the decision-maker's preference relation be a strict partial ordering and let $R \neq \varnothing$. Suppose for some subset T of the criteria set K, T is strictly independent of its complement, $K-T$. Let us assume that the conclusion of the theorem is false and $K-T$ is also strictly independent of T.

Since $R \neq \varnothing$, $\exists \mathbf{x}, \mathbf{y} \in \mathbb{R}^n$ such that $\mathbf{x}R\mathbf{y}$. Consider $\mathbf{u}^{(1)}, \mathbf{w}^{(1)} \in \mathbb{R}^n$ such that $\mathbf{x}|_T = \mathbf{u}^{(1)}|_T$ and $\mathbf{y}|_T = \mathbf{w}^{(1)}|_T$; since T is strictly independent of $K-T$, it follows that $\mathbf{u}^{(1)}R\mathbf{w}^{(1)}$. Next, consider $\mathbf{u}^{(2)}, \mathbf{w}^{(2)} \in \mathbb{R}^n$ such that $\mathbf{u}^{(1)}|_{K-T} = \mathbf{u}^{(2)}|_{K-T}$ and $\mathbf{w}^{(1)}|_{K-T} = \mathbf{w}^{(2)}|_{K-T}$, and $\mathbf{u}^{(2)}|_T = \mathbf{y}|_T$ and $\mathbf{w}^{(2)}|_T = \mathbf{x}|_T$; since $K-T$ is strictly independent of T, it follows that $\mathbf{u}^{(2)}R\mathbf{w}^{(2)}$. Finally, consider $\mathbf{u}^{(3)}, \mathbf{w}^{(3)} \in \mathbb{R}^n$ such that $\mathbf{u}^{(3)}|_T = \mathbf{u}^{(2)}|_T$ and $\mathbf{w}^{(3)}|_T = \mathbf{w}^{(2)}|_T$, and $\mathbf{u}^{(3)}|_{K-T} = \mathbf{y}|_{K-T}$ and $\mathbf{w}^{(3)}|_{K-T} = \mathbf{x}|_{K-T}$; since T is strictly independent of $K-T$, it follows that $\mathbf{u}^{(3)}R\mathbf{w}^{(3)}$. But, by construction, $\mathbf{u}^{(3)} = \mathbf{y}$ and $\mathbf{w}^{(3)} = \mathbf{x}$. So $\mathbf{y}R\mathbf{x}$. Therefore, we have $\mathbf{x}R\mathbf{y}$ and $\mathbf{y}R\mathbf{x}$. This is a contradiction to R being a strict partial ordering. Hence, if T is strictly independent of $K-T$, then $K-T$ is not strictly independent of T, and relation of strict independence is asymmetric. *Q.E.D.*

Theorems 4.1, 4.2, and 4.4 and Corollary 4.3 clarify the concepts of preferential independence on one hand and of strict independence on the other. The concept of strict independence establishes an asymmetric relation among criteria sets. This concept contrasts the idea of mutual preferential independence, an equivalence relation on the criteria set, which was used by Keeney and Raiffa and others to lay a foundation for additive value functions. The results also help us develop an effective approach to treating qualitative criteria based on relative importance of criteria groups.

4.4 A SEQUENCE OF "RELAXED" MULTICRITERIA PROBLEMS AS A WAY OF TREATING QUALITATIVE CRITERIA SUBSETS

Consider a multiple criteria problem with $\mathbb{R}^m$ as the decision space containing alternative solutions as vectors of m parameter values. Let $\tilde{A}$ be the set of feasible solutions in the decision space: $\tilde{A} \subseteq \mathbb{R}^m$. Suppose that a subset T of the criteria set K, $T = \{K_{i_1}, K_{i_2}, \ldots K_{i_r}\}$, is preferentially independent of its complementary set $K-T$; let $\mathbb{R}^r$ be the space containing vectors of ordered evaluations of alternative solutions with respect to the criteria set T, and let $\mathbf{f}: \tilde{A} \to \mathbb{R}^r$ be the criteria mapping, which assigns to each element $\mathbf{a} \in \tilde{A}$ a vector of r criteria values $\mathbf{f}(\mathbf{a}) \in \mathbb{R}^r$. Let $A = \mathbf{f}(\tilde{A})$ be the image of $\tilde{A}$ under $\mathbf{f}$. Assume also that there exists preference structure in $\mathbb{R}^r$: a choice function $C(\cdot)$.

Definition 4.4.1

For a given subset T of the criteria set K, $T=\{K_{i_1}, K_{i_2}, \ldots K_{i_r}\}$, the corresponding "relaxed" multicriteria problem is defined as the problem of choosing optimal solutions with respect to the set T only, without regard of the complementary criteria set. An optimal solution for the relaxed problem chosen from a subset of feasible solutions $Y \subseteq \widetilde{A}$ is defined as any element $\mathbf{y} \in Y$ such as $\mathbf{y} \in \mathbf{f}^{-1}(C(\mathbf{f}(Y)))$.

Suppose a decision analyst determines that the decision maker has several criteria to base his or her decision on, while a subset of the criteria set consists of qualitative, hard-to-measure goals and objectives. Often, such criteria characterize intangible aspects of the solutions. Let us consider two major cases. In case one, a qualitative criterion is important in the sense that its value must satisfy a minimum requirement without which any solution is prohibited. For example, in a hiring problem, best candidates must first satisfy such minimum requirements as academic degree and minimum quality of previous work experience, which would exemplify case one. In case two, a qualitative criterion is important but less so than other, measurable objectives. For instance, in the hiring problem the overall fit with the organization would probably represent case two. Both cases are tied to relative importance of qualitative versus quantitative criteria; depending on a case, we will treat qualitative criteria differently as part of our multi-criteria model.

Based on these classification of qualitative criteria, let us assume that the set of N criteria $\{K_1, K_2, \ldots, K_N\}$ is partitioned into three subsets: (1) the set P of p qualitative criteria that take precedence over all other criteria; (2) the set S of all s quantitative criteria that represent tangible values of alternative solutions; (3) the set Q of q qualitative criteria that represent intangibles of lesser importance compared to sets P and S. Relative importance of the sets P, S, and Q will be understood in the sense of Definition 4.3.9. Given this structure of the criteria set K, we propose a sequential approach to constructing a multi-criteria optimization model. This approach is expressed as the following three-phase procedure.

Phase I: Let $\mathbb{R}^m$ be the decision space: the set containing alternative solutions as vectors of m parameter values. Let $\widetilde{A}$ be the set of feasible solutions in the decision space: $\widetilde{A} \subseteq \mathbb{R}^m$. Given criteria group P, Phase I creates a modified set of feasible solutions $\widehat{A} \subseteq \mathbb{R}^m$ where $\widehat{A} = \{\mathbf{a} \in \widetilde{A} \mid \mathbf{a} \text{ satisfies minimum requirements for set } P\}$.

Phase II: Given the set of quantitative criteria $S=\{K_{i_1}, K_{i_2}, \ldots K_{i_s}\}$, solve the corresponding "relaxed" multicriteria problem in criteria space $\mathbb{R}^s$. Given a criteria mapping $\mathbf{f}\colon \widetilde{A} \rightarrow \mathbb{R}^s$, an optimal solution for the "relaxed" problem chosen from a subset of feasible solutions $Y \subseteq \widehat{A}$ is defined as $M(Y)=\mathbf{f}^{-1}(C(\mathbf{f}(Y)))$.

Phase III: Given the set of qualitative criteria Q, present it to the decision maker and apply multicriteria methodology as necessary to solve the corresponding "relaxed" multi-criteria problem in criteria space $\mathbb{R}^q$. An optimal solution for this "relaxed" problem is chosen from $M(Y)$, the output of Phase II.

Phases II and III represent solving "relaxed" multi-criteria problems, which are much less complex than the initial criteria space. Note that each (but not all) of the criteria sets P, S, and Q may be empty; then the corresponding phase is eliminated.

Applying the three-phase process formulated here allows fulfilling the following objectives: decomposing of the criteria set into more uniform groups; separating intangible, qualitative criteria from tangible, quantitative criteria without artificially diminishing the role of qualitative criteria; creating possibilities of working out an efficient optimization model in the space of quantitative criteria during Phase II; and giving to the decision maker a bigger role in completing the final selection of optimal solutions during Phase III. Note that Phase II need not be interactive if preferences are known in advance. It is conceivable that the first two phases are completed once while Phase III is repeated as necessary in real time to address the decision-maker's changing and sometimes hard to formalize preferences. In addition, Phase II does not have to result in a single optimal solution because Phase III allows for further narrowing the set; therefore, it is possible to carefully reveal decision-maker's preferences without the necessity of settling on a scalarized model such as value function. When some alternative solutions may not be comparable for the decision maker, binary preference relations at both second and third phases of the process appear to be viable alternatives to scalarized models.

4.5 CONCLUSIONS

We started this chapter by stating some methodological and philosophical challenges when facing intangible, qualitative objectives in multi-criteria decision making. Although tangible, quantitative criteria can be objectively evaluated, qualitative criteria by their nature forbid assignment of objective values. Inevitably, attempts to convert qualitative evaluations into numerical form bring elements of subjectivity into the evaluation process and incorporate preferences in it. This disrupts the multiple-criteria optimization model with its three components: decision space, criteria space, and criteria mapping because then elements of the criteria mapping are influenced by preferences. Ideally, the decision-making process occurs in the criteria space where decision-maker's preferences are defined, while decision space, criteria space, and criteria mapping are based in objective reality (Buchanan et al., 1998). Separating qualitative criteria from quantitative helps eliminate bias from the part of decision making dealing with the measurable, quantitative criteria. Objective and expert information on the decision-maker's tangible goals is represented by Phase II where we solve a "relaxed" multi-criteria problem without the interference of qualitative, intangible values. The set of solutions selected at Phase II is then evaluated with respect to the remaining qualitative criteria in Phase III, another "relaxed" multi-criteria problem. At this point, building preference models might involve a higher level of participation from the decision maker because intangible goals are elusive, subjective, and hard to formalize.

Ultimately, the approach proposed herein gives several advantages over the idea of using a single multi-criteria space. Firstly, it allows to separate objective and expert information addressing the decision-maker's goals from informal, subjective views of both the experts and the decision maker. Secondly, this approach allows for the developing of more efficient ways of modeling preferences in the "relaxed" problem with all-quantitative criteria. Such methodologies as trade-offs, gradient search, and value functions are more viable in such setting. Finally, this approach allows to fully address all decision-maker's objectives regardless of their nature.

Although this chapter is an attempt to find ways of treating qualitative criteria efficiently, we did not set a task here of evaluating such criteria or creating multi-criteria methods in the space of all-qualitative criteria. These tasks are better addressed when they are intended for a specific application at hand.

REFERENCES

Arrow, K. J. (1959). Rational choice functions and orderings, *Economica, New Series*, 26(102), 121–127.

Alfares, H. K., & Duffuaa, S. O. (2009). Assigning cardinal weights in multi-criteria decision making based on ordinal ranking, *Journal of Multi-criteria Decision Analysis*, 15, 125–133.

Belton, V., & Stewart, T. J. (2002). *Multiple Criteria Decision Analysis*. Kluwer, New York.

Buchanan, J. T., Henig, E. J., & Henig, M. I. (1998). Objectivity and subjectivity in the decision making process, *Annals of Operations Research*, 80, 333–345.

Doumpos, M., Zanakis, S. H., & Zopounidis, C. (2001). Multicriteria preference disaggregation for classification problems with an application to global investing risk, *Decision Sciences*, 32, 333–385.

Keeney, R. L., & Raiffa, H. (1976). *Decisions with Multiple Objectives*. Wiley: New York.

Kostreva, M. M., & Polyashuk, M. V. (1993). Resolution of dilemmas: A mathematical theory, *Journal of Multi-criteria Decision Analysis*, 2, 159–166.

Markowitz, H. M. (1959). *Portfolio Selection: Efficient Diversification of Investments*. Wiley: New York.

Plott, C. R. (1976). Axiomatic social choice theory: An overview and interpretation, *American Journal of Political Science*, 20(3), 511–596.

Polyashuk, M. V. (2005). A formulation of portfolio selection problem with multiple criteria, *Journal of Multi-criteria Decision Analysis*, 13, 135–145.

Savage, L. J. (1972). *The Foundations of Statistics*. Dover Publications, New York.

Sen, A. K. (1971). Choice functions and revealed preference, *The Review of Economic Studies*, 38(3), 307–317.

Zeleny, M. (2011). Multiple criteria decision making (MCDM): From paradigm lost to paradigm regained? *Journal of Multi-criteria Decision Analysis*, 18, 77–89.

CHAPTER 5

Some Methods and Algorithms for Constructing Smart-City Rankings

Esther Dopazo and María L. Martínez-Céspedes

CONTENTS

5.1 INTRODUCTION

City rankings have become central instruments for assessing the attractiveness of urban regions over the last 20 years [1–3]. Demographic, environmental, economic, political, and sociocultural factors are forcing the urban world to design and implement *smart cities*. Smart-city measures are achieved through carefully chosen indicators and allow cities to reorganize itself successfully via an understanding of their strengths and weaknesses. In comparative studies, cities are evaluated and ranked in regard to different economic, social, environmental, and geographical evaluation criteria to reveal the best places for certain activities [1–9]. The number of rankings in existence around the world is growing year after year. Therefore, the problem of combining multiple rankings to form a synthetic ranking, which compare city performance, is recognized as a useful tool in policy analysis, city marketing, benchmarking, and public communication.

The problem of rank a set of cities attending their smart-city nature can be considered as a multi-criteria problem [4,5] because of the multidimensionality character of the smart-city concept [1–3]. The principal aim is to aggregate and exploit data information from several ranked lists according to different criteria to produce a synthetic smart-city ranking, as it is illustrated in Figure 5.1.

Synthetic indexes are common in fields such as economic and business statistics (e.g., the OECD Composite of Leading Indicators by Nardo [10]) and they are used in a variety of policy domains. The proliferation of these indexes is a clear symptom of their political importance and operational relevance in management decision making.

The problem of rank aggregation is not new. It has been studied extensively since the eighteenth century when Borda and Condorcet [11,12] published their works on Social Choice Theory. Kemeny-Snell [13] suggested

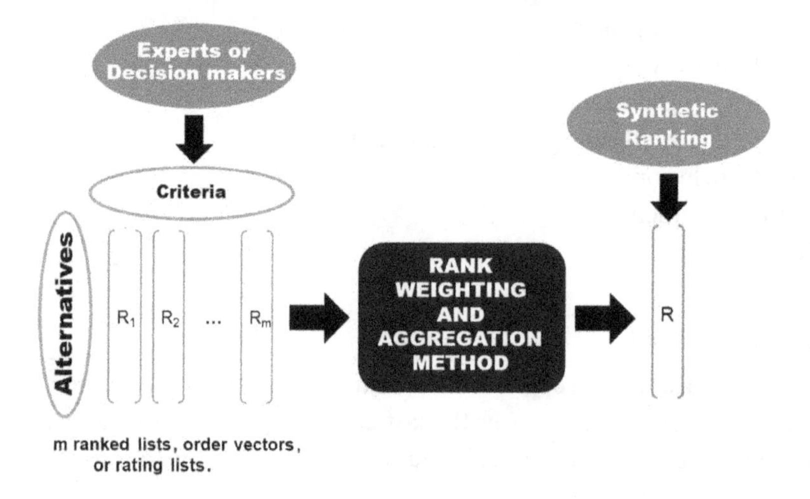

FIGURE 5.1 Rank aggregation overview.

determining the aggregate ranking that minimizes the sum of deviations from the individual rankings. However, this optimization problem was proven to be NP-hard [14]. Cook and Seiford [15] provided a consensus ranking in terms of the median ranking. Distance-based methods dealing with ranking aggregation are proposed in [16]. More recently, the problem has been addressed in specific contexts that pose significant challenges for rank aggregation like Web metasearch engines [14,17], information retrieval [18,19], sports [20], environment sustainability [4], and smart cities [2,3,21]. In [22] and in [23] the problem is studied dealing with uncertain data. A great deal of the literature is concerned with methods that assume full lists. These models cannot be applied directly in the context of smart cities, where partial indicators seems to be common. Then, a challenge in this context is to provide methods and algorithms to generate aggregate rankings able to deal with limited, conflicting, and heterogeneous information from different sources [21,24].

The motivation and scope of this chapter is to extend methods in [21] (to deal with multiplicative consistency) to determine an aggregate ranking and to illustrate their application using data from the report *Cities in Motion Index 2016* (*CIMI* 2016) [25] provided by the Centre of Globalization and Strategy of IESE Research Center. The proposed methods are articulated in two steps. In the first step an outranking matrix [5,26] is constructed according all the individual rankings as a way of summarizing ordinal information. This matrix expresses dominance between pairs of cities. Then, the problem is to exploit information collected in this matrix to provide a synthetic ranking. In the first method, priority values are calculated using data-fitting techniques following a multiplicative fuzzy preference model [27]. In the second method, a procedure similar to Google PageRank algorithm [28] is used. Finally, the items are ranked in decreasing order attending to the priority values.

The proposed approach provides a theoretical framework and computational tools to address the rank aggregation problem dealing with ordinal data. The first step gives us a way to deal with ordinal rankings without treating and operating with them as if they were precise real valued vectors [16]. Moreover, requirements of full lists and homogeneity of criteria are relaxed. Finally, the methods derive priority vectors of the considered items; that it is more desirable than a single-rank position list, providing more information to analyze the final ranking.

The rest of the chapter is organized as follows. In Section 5.2, some rank aggregation methods and algorithms are presented. In Section 5.3, the proposed analytic framework is contrasted and applied using a data set provided by *Cities in Motion Index 2016* report [25]. Finally, main conclusions derived from this research are outlined in Section 5.4.

5.2 RANK AGGREGATION METHODS

The rank aggregation problem, also known as group-ranking problem, consists of combining several ranked lists of a set of alternatives in a robust way to produce a synthetic ranking (Figure 5.1). This synthetic ranking summarizes input information on the alternatives. It is recognized as a useful tool in several fields including meta-search engines, recommendation systems, information retrieval, multi-criteria decision making, and social choice theory.

Let $X = \{x_1, x_2, \dots, x_n\}$ $(n \geq 2)$ be the set of items (cities in our case) that have to be ranked according to information contained in the ordinal rankings $R^k = (r_i^k)$, $k = 1, \dots, m$ $(m > 1)$, of the items with respect to m criteria. r_i^k denotes the position or rank of item x_i (a highly ranked element has a low ordinal position) with respect to the k-criterion, where $i = 1, \dots, n$, and $k = 1, \dots, m$. Let $w = (w_1, \dots, w_m)$ be the weight vector of the criteria, where $w_k > 0$ represents the relative importance assigned to the k-th criterion and $\sum_{k=1}^{m} w_k = 1$.

The vector w is usually provided by the manager or organizer. If vector w has not been stated a priori, the AHP method [29], a leading multicriteria decision-making method, is a possible option for deriving this vector. In AHP, the decision problem is structured as a hierarchy of criteria and subcriteria. Then a pairwise comparison matrix $W = (w_{ij})$ is constructed, where w_{ij} reflects the importance ratio between the criteria i and criteria j according to the Saaty scale [29]. The relative weights $w_1, \dots, w_m$ are calculated from this matrix W using the eigenvector method [29] or a distance-based method [30,31].

We focus on constructing a synthetic ranking on X, which best reflects the information expressed by means of individual ordinal rankings R^k, $k = 1, \dots, m$. We propose a two-stage approach to derive a synthetic ranking of the alternatives. In the first stage, we create an aggregate dominance matrix as a model for gathering preferences from multiple input rankings. In the second stage, we consider two methods for deriving a priority vector from the aggregate dominance matrix based on properties of fuzzy preference relations and graph theory. Finally, the alternatives are ranked in decreasing order according to the obtained priority values.

First we construct the dominance matrix $X^k = (x_{ij}^k)$ associated to the k-th criteria as follows:

$$x_{ij}^k = \begin{cases} 1 & \text{if item } x_i \text{ is ranked above item } x_j, r_i^k < r_j^k, \\ 1/2 & \text{if item } x_i \text{ is tied with item } x_j, r_i^k = r_j^k, \\ 0 & \text{if item } x_i \text{ is ranked below item } x_j, r_i^k > r_j^k, \end{cases} \quad (5.1)$$

where $i, j = 1, \ldots, n$, and $k = 1, \ldots, m$.

5.2.1 Rank Aggregation Method 1: Fuzzy Preference Relation Approach

We first create the $n \times n$ aggregate dominance matrix $A = (a_{ij})$, where a_{ij} quantifies the extent to which alternative x_i is preferred to alternative x_j according to the input rankings $R^k, k = 1, \ldots, m$, and taking into account the weight criteria vector w:

$$a_{ij} = \sum_{k=1}^{m} w_k x_{ij}^k, \qquad i, j = 1, \ldots, n. \quad (5.2)$$

Notice that $a_{ij} \in [0,1]$ expresses the relative preference degree of alternative x_i over alternative x_j. Larger value of a_{ij} denotes a greater preference and a lower ordinal position in the ranking. Thus, the unit interval scale has the following associated semantic:

$$a_{ij} = \begin{cases} 1 & \text{if } x_i \text{ is definitely preferred over } x_j, \\ (0.5, 1) & \text{if } x_i \text{ is preferred over } x_j, \\ 0.5 & \text{if there is indifference between } x_i \text{ and } x_j, \\ (0, 0.5) & \text{if } x_j \text{ is preferred over } x_i, \\ 0 & \text{if } x_j \text{ is definitely preferred over } x_i. \end{cases} \quad (5.3)$$

Furthermore, matrix A satisfies the reciprocity property $a_{ij} + a_{ji} = 1$, $i, j = 1, \ldots, n$ [27]. It can be considered that the binary relation defined in A describes a fuzzy preference relation [27,32] summarizing preference information in R^k, $k = 1, \ldots, m$. A similar matrix has been used in [4], and in [22] and [23] dealing with interval ordinal data.

Once the matrix A has been created, we focus on deriving a positive priority vector $v = (v_1, \ldots, v_n)^t$ where v_i reflects the priority degree of the alternative (item) x_i from the fuzzy preference relation expressed in matrix A. In the ideal case, where A is multiplicative consistent [27,32], this is

$$\frac{a_{ij}}{a_{ji}}\frac{a_{jk}}{a_{kj}}=\frac{a_{ik}}{a_{ki}}, \qquad i,j,k=1,\ldots,n,$$

then there exists a positive priority vector $v=(v_i)$ such that

$$a_{ij}=\frac{v_i}{v_i+v_j}, \qquad i,j=1,\ldots,n, \tag{5.4}$$

with the usual normalization condition $\sum_{i=1}^{n} v_i=1$.

However, due to the complexity of real-world problems matrix A does not, in the general case, satisfy the consistency requirements [27]. Hence the following overdetermined linear system from (4) has no solution:

$$\begin{cases}(1-a_{ij})v_i-a_{ij}v_j, & i,j=1,\ldots,n,\\ \sum_{i=1}^{n} v_i=1. \end{cases} \tag{5.5}$$

Then, the problem could be formulated to find a rank vector v or the associated transitive matrix $B=(b_{ij})$, where $b_{ij}\in\{0,1/2,1\}$, that best approximates matrix A ($\|A-B\|$ be minimal for some matrix norm $\|\,\|$).

Therefore, we focus on deriving a normalized positive priority vector $v=(v_1,\ldots,v_n)$ to find the best solution to the overdetermined linear system (5), in the sense that minimizes deviations $\delta_{ij}:=|(1-a_{ij})v_i-a_{ij}v_j|$, $i,j=1,\ldots,n$, under a usual l_p norm [33]. This is, v is a positive l_p-solution of system (5) in the following sense (least square approximation):

$$Min\left\{\sum_{i,j=1}^{n}((1-a_{ij})v_i-a_{ij}v_j)^2\right\}, \tag{5.6}$$

subject to $v_i>0, i,j=1,\ldots,n$, and $\sum_{i=1}^{n} v_i=1$. Since matrix A is reciprocal ($a_{ij}+a_{ji}=1; \forall i,j$), in the preceding optimization problem it is enough to consider a_{ij} where $i<j$. The optimization problem (5.6) results into a least squares problem for which many numerical tools are available to compute the solution. This approximation approach has been considered by [23] dealing with uncertain data.

Once the priority vector $v=(v_i)$ has been computed we output the final ranking $R=(r_i)$ of items $x_i, i=1,\ldots,n$, by sorting them in descending order according to the values $v_i, i=1,\ldots,n$.

5.2.2 Rank Aggregation Method 2: Perron Eigenvector Approach

The second method follows the ideas of Keener's rating method [20,34] and the PageRank algorithm [28]. We consider the preference matrix $A=(a_{ij})$ from (2), where a_{ij} quantifies the extent to which alternative x_i is preferred to alternative x_j according to the input rankings. Moreover, it would be worthwhile, when comparing x_i with x_j, to scale the dominance of x_i over x_j (expressed by a_{ij}) to the total dominance of the alternatives over x_j $\left(\sum_{l=1}^{n} a_{lj}\right)$. For instance, the scaled dominance of x_i over x_j is given by

$$s_{ij}=\frac{a_{ij}}{\sum_{l=1}^{n} a_{lj}}, \qquad i,j=1,\ldots,n. \tag{5.7}$$

Matrix $S=(s_{ij})$ describes a weighted digraph, where the set of nodes is the set X of alternatives. s_{ij} is the weight attached to the directed arc (x_i, x_j) denoting the relative paired comparisons of the strengths of the alternatives. It can be interpreted as the probability (relative strength) of alternative i dominating alternative j, and the normalized-column matrix S as the transition matrix of the weighted digraph. Accordingly, the j-column in S represents the probability vector of dominance over alternative j. Matrix S is a sub-stochastic matrix: $s_{ij}\geq 0$ and $\sum_{i=1}^{n} s_{ij}\leq 1$. The sum $\sum_{i=1}^{n} s_{ij}=1$ excepting $s_{ij}=0$, $\forall i\neq j$. This case corresponds with the situation where $a_{ij}=0\ \forall i\neq j$, that means $x_{ij}^{k}=0\ \forall i\neq j$, $k=1,\ldots,m$ (2), or $r_i^k > r_j^k\ \forall i\neq j, k=1,\ldots,m$. Therefore, it corresponds with the case x_i is ranked below or dominated by $x_j\ \forall i\neq j$, $k=1,\ldots,m$.

Now, we look for a normalized nonnegative rating vector $r=(r_1,\ldots,r_n)^t$ $\left(\|r\|_1=\sum_{i=1}^{n} r_i=1\right)$, where r_i indicates the importance of x_i based on relative paired comparisons expressed in matrix S. As suggested by Keener's rating method [20,34], the importance r_i of alternative x_i is estimated from its interactions with the other alternatives along one criteria or ranking. Therefore, the rating or overall importance r_i of x_i is considered to be the sum of the relative importances of alternative x_i with respect to all other alternatives:

$$r_i=\sum_{j=1}^{n} s_{ij} r_j, \qquad i=1,\ldots,n. \tag{5.8}$$

The above equation can be written in matrix form: $r = Sr$.

Then the rating vector, r, is given by a positive eigenvector associated to the Perron eigenvalue of matrix S [34]. In practice, the power method, a simple iterative algorithm (basically, $Sr^{k+1} = \lambda r^k$), can be used to obtain an approximation of the rating vector, r.

This is equivalent to saying that r is a nonnegative eigenvector associated with the eigenvalue 1 (Perron eigenvalue) of matrix S [34], subject to the normalization condition $\|r\|_1 = 1$. In practice, the power method [35] is a simple iterative algorithm to compute r, where, starting with an arbitrary initial vector r^0 with $r^0 > 0$ and $\|r^0\|_1 = 1$, the normalized sequence r^k $\left(\|r^k\|_1 = 1\right)$ is calculated as follows:

$$r^{k+1} = Sr^k, \qquad k = 0,\ldots,m. \tag{5.9}$$

The sequence $\{r^k\}$ provides approximations of the importance vector r.

5.3 ILLUSTRATIVE CASE STUDY

In this section, we illustrate the application of the proposed methods using the data provided by the report *CIMI* 2016 [25], conducted by the Centre of Globalization and Strategy of IESE Research Center. CIMI provides a solid understanding of the way cities operates to promote economic development and to improve competitiveness. The prestigious report *Business of Cities 2015* by JLL (Jones Lang Lasalle, see [2]) considered CIMI 2014 edition as the best city index in the world in the areas of planning, strategy, and innovation capacity of cities.

The *CIMI* 2016 edition involves 181 cities representing more than 80 countries. It evaluates cities' level of development in relation to 10 key criteria or dimensions $C^1,\ldots,C^{10}$: Economy, Human Capital, Social Cohesion, Environment, Public Management, Governance, Urban Planning, International Outreach, Technology, and Mobility and Transportation. The *CIMI* 2016 study provides the criteria weight vector $w = (w_i)$ shown in Table 5.1 obtained by using an iterative procedure called DP2 [36,37] taking into account dependencies of the indicators. This weight vector, w, is normalized using the vectorial norm l_∞ $\left(\|w\|_\infty = \max_{i=1,\ldots,10}\{|w_i|\} = 1\right)$. For our purpose, to assure that the dominance values expressed in matrix A are in the interval $[0,1]$, we

TABLE 5.1 *CIMI 2016* Criteria Relative Weights

Id	Criteria	Weight
C^1	Economy	1.0000
C^2	Human Capital	0.4814
C^3	Social Cohesion	0.5941
C^4	Environment	0.6215
C^5	Public Management	0.5710
C^6	Governance	0.4070
C^7	Urban Planning	0.8410
C^8	International Outreach	0.6212
C^9	Technology	0.3763
C^{10}	Mobility and Transportation	0.4707

normalize this weight vector using the vectorial norm l_1, that means $\| w \|_1 = \sum_{i=1}^{10} | w_i | = 1$. It produces de weight vector:

$$w = (0.1671, 0.0804, 0.0993, 0.1039, 0.0954, \\ 0.068, 0.1405, 0.1038, 0.0629, 0.0787). \tag{5.10}$$

This normalization condition does not affect to the final ranking.

In Figure 5.2 is highlighted the relevance of the criteria. In the general case where the criteria weights have not been stated a priori, we propose to use a multi-criteria decision making method, for instance the analytic hierarchy process (AHP) [29], to derive the criteria weight vector based on the information provided by a panel of experts.

5.3.1 Data Set

To illustrate the application of the methods and just for sake of simplicity, we consider a sample of cities studied in the *CIMI 2016* [25]. Let X be the set of cities classified as high performance cities by the *CIMI* study, that are cities whose CIMI performance value is greater or equal than 90. It results $X = \{x_1, x_2, \ldots, x_7\}$ where $x_1 =$ New York, $x_2 =$ London, $x_3 =$ Paris, $x_4 =$ San Francisco, $x_5 =$ Boston, $x_6 =$ Amsterdam, and $x_7 =$ Chicago.

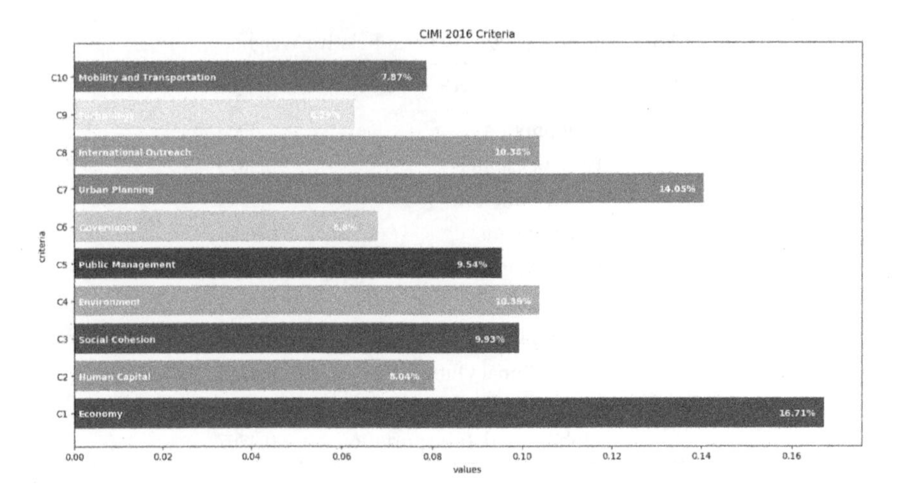

FIGURE 5.2 Criteria weights.

Let's assume that the ten dimensions or criteria $C^1,\ldots,C^{10}$ are considered where vector w (5.10) expresses the criteria weight vector following *CIMI* study [25]. Now, let's take into consideration the ordinal data in the rank orderings $R^k = (r_i^k)$, $i = 1,\ldots,7$, $k = 1,\ldots,10$ (Table 5.2) of the studied cities derived from the partial indicators representative of the 10 criteria in the *CIMI* study [25].

5.3.2 Method Application and Results

Now, we focus on applying the proposed methods to rank the considered cities according to the information contained in the ordinal rankings $R^k = (r_i^k)$, $k = 1,\ldots,10$, of cities with respect to the considered 10 criteria

TABLE 5.2 Partial Rankings According to Considered Criteria

City	C_1	C_2	C_3	C_4	C_5	C_6	C_7	C_8	C_9	C_{10}	Borda Count
New York	1	3	6	6	1	1	6	3	1	2	30
London	3	1	6	6	1	1	6	2	3	1	30
Paris	6	4	4	2	5	5	3	1	7	3	39
San Francisco	2	6	3	5	4	3	4	6	5	7	45
Boston	5	2	1	3	2	1	5	7	6	4	36
Amsterdam	7	7	2	1	6	4	1	4	2	6	40
Chicago	4	5	5	4	3	1	2	5	4	5	38

Source: *CIMI 2016.*

(Table 5.2) and their attached weights (5.10). We first create the 7×7 aggregate dominance matrix $A=(a_{ij})$ according to (5.2):

$$A=\begin{pmatrix} 0.5 & 0.4835 & 0.5525 & 0.6563 & 0.5418 & 0.6563 & 0.6223 \\ 0.5165 & 0.5 & 0.5525 & 0.4892 & 0.6223 & 0.59344 & 0.6223 \\ 0.4475 & 0.4475 & 0.5 & 0.5073 & 0.4268 & 0.5254 & 0.4660 \\ 0.3437 & 0.5107 & 0.4928 & 0.5 & 0.4743 & 0.4109 & 0.2664 \\ 0.4581 & 0.3776 & 0.5731 & 0.5256 & 0.5 & 0.5889 & 0.4916 \\ 0.3436 & 0.4065 & 0.4745 & 0.5890 & 0.4111 & 0.5 & 0.510 \\ 0.3776 & 0.3776 & 0.5339 & 0.7336 & 0.5083 & 0.4896 & 0.5 \end{pmatrix}.$$

For instance, the figure a_{12} quantifies the extend to which x_1 (New York) dominates x_2 (London) according to $R^k=(r_i^k)$, $k=1,\ldots,10$. From data in Table 5.2 we can see that New York dominates London for criteria C^1 and C^9 $\left(x_{12}^1=x_{12}^9=1\right)$ and gets tied for criteria C^3, C^4, C^5, C^6, C^7 $\left(x_{12}^3=x_{12}^4=x_{12}^5=x_{12}^6=x_{12}^7=0.5\right)$. Then, these dominances and ties are quantified following (5.2), attending to the criteria weights (10) as follows:

$$a_{12}=w_1+w_9+0.5\left(\sum_{i=3}^{7} w_i\right).$$

Notice that A represents a fuzzy preference relation, where $a_{ij}\in[0,1]$ indicates the preference degree of alternative x_i over x_j. Moreover, matrix A satisfies the reciprocity property $a_{ij}+a_{ji}=1$, $i,j=1,\ldots,7$.

Following *Method 1* described in Section 5.2.1 and solving the resulting least squares problem (5.6) using MATLAB numerical software, we obtain the priority vector v = (0.1896, 0.1792, 0.1306, 0.1038, 0.1422, 0.1199, 0.1347) from the fuzzy preference relation A. It produces the following rank ordering:

New York ≻ London ≻ Boston ≻ Chicago ≻ Paris ≻ Amsterdam ≻ San Francisco.

To apply the method in Section 5.2.2, we first construct the normalized-column matrix S according to (5.7) from A:

$$S=\begin{pmatrix} 0.1674 & 0.1558 & 0.1502 & 0.1640 & 0.1555 & 0.1743 & 0.1789 \\ 0.1729 & 0.1611 & 0.1502 & 0.1223 & 0.1786 & 0.1576 & 0.1789 \\ 0.1498 & 0.1442 & 0.1359 & 0.1268 & 0.1225 & 0.1396 & 0.1340 \\ 0.1151 & 0.1646 & 0.1339 & 0.1250 & 0.1361 & 0.1092 & 0.0766 \\ 0.1534 & 0.1217 & 0.1558 & 0.1314 & 0.1435 & 0.1564 & 0.1413 \\ 0.1151 & 0.1310 & 0.1290 & 0.1472 & 0.1180 & 0.1328 & 0.1467 \\ 0.1264 & 0.1217 & 0.1451 & 0.1834 & 0.1459 & 0.1301 & 0.1437 \end{pmatrix}.$$

Notice that S is a stochastic matrix ($s_{ij} \geq 0$ and $\sum_{i=1}^{7} s_{ij} = 1$). We now compute the rating vector $r = (0.1636, 0.1613, 0.1367, 0.1237, 0.1432, 0.1307, 0.1410)^t$ as the normalized positive eigenvector of matrix S associated to the dominant eigenvalue (that is 1). It provides the following rank ordering:

New York ≻ London ≻ Boston ≻ Chicago ≻ Paris ≻ Amsterdam ≻ San Francisco.

The top-7 CIMI ranking and the results yielded by our approach are summarized in Table 5.3. We conclude that the proposed methods provide the same two top cities than CIMI ranking. We notice that they confirm that New York followed by London are the best alternatives as in *CIMI* 2016 index.

Related to Boston, Chicago, and Paris we note that the priority values derived from our methods are very close. The priority values derived by our methods of these three cities are very close, even more the corresponding

TABLE 5.3 Rank Orderings

Ranking	CIMI 2016	Method 1	Method 2
1	New York	New York	New York
2	London	London	London
3	Paris	Boston	Boston
4	San Francisco	Chicago	Chicago
5	Boston	Paris	Paris
6	Amsterdam	Amsterdam	Amsterdam
7	Chicago	San Francisco	San Francisco

values in the preference matrix A are close to 0.5, meaning they are almost tied. We observe that Boston is ranked higher than Paris according five criteria, being one of them the more relevant criteria, C^1. Boston and Paris are the best options respect to criteria C^3 and C^8, respectively, with quite similar weights. Boston is ranked in the third position and Paris in the fifth position from the calculated priority values $v_5 = 0.1422$ and $v_3 = 0.1306$, attending to the first method, and $r_5 = 0.1432$ and $r_3 = 0.1367$, attending to the second method. In both cases they reveal deviations about or less than 8%. Proceeding in the same way, the deviations of the priority values of Chicago and Paris (fourth and fifth position respectively in the proposed ranking) are at most of the 3%.

On the other hand, San Francisco occupies the last position attending to our methods. We notice that all the preference values in the fourth row of matrix A are less than 0.5, but only one is close to 0.5. They represents the performance of San Francisco respect to the other cities along the criteria. If we consider the sum of the fourth row of A, we can see that produces the lowest value with the respect to the sum of the other rows in matrix A. We notice that San Francisco is never considered the best option along the criteria as the other cities are. Using the Borda count, consisting on adding the individual marks in Table 5.2 for each city, San Francisco occupies the worst position.

5.4 CONCLUSIONS

In this chapter, we have proposed two flexible rank aggregation methods to produce a synthetic ranking that summarize information from individual ordinal rankings. It addresses some of the shortcomings that appear in this context.

- The proposed approach provides a mathematical foundation to study the problem as opposite of heuristic methods used in some applications. Moreover, computational procedures can be used for solving the problem (least squares procedure and power method).
- The proposed methods can handle partial lists and rankings that contain ties.
- In the first stage of our approach, a pairwise matrix A (5.2) is constructed to express dominance relations as a way to avoid to deal with the ordinal rank positions as if they were precise real values, as in some methods in the literature.

- A numerical example illustrates the application of the methods by using public data from the IESE *Cities In Motion Index* 2016 (*CIMI* 2016) report [25]. There is a lot of research in select the adequate metrics to measure the smart-city dimensions, also in the metadata structure, but there is a lack of research in how to aggregate this information through an established model.

More specifically, this work extends previous studies and presents a methodology to construct a consensus ranking in virtue of the idea of rank aggregation methods using selected and available data from smart cities. An objective for future work is to extend these methods to deal with uncertain data expressed in the form of (user-friendly) linguistic preferences and fuzzy numbers and to provide an integrated model dealing with ordinal and cardinal data.

ACKNOWLEDGMENTS

The authors are grateful to the reviewers and editor for their valuable and constructive comments and suggestions. In addition, the authors are grateful IBM for investing in the original concept of this work through its prize donation under the umbrella of the Smart City Planet comes to you Challenge, 2013.

REFERENCES

1. T. Moonen, G. Clark, The business of cities 2013, available: http://www.jll.com/Research/jll-city-indices-november-2013.pdf (2013).
2. G. Clark, T. Moonen, J. Couturier, *The Business of Cities 2015* (2015).
3. R. Giffinger, C. Fertner, H. Kramar, R. Kalasek, N. Pichler-Milanovic, E. Meijers, Smart cities—Ranking of European medium-sized cities, final Report, Centre of Regional Science, Vienna UT, Vienna, Austria (2013).
4. G. Munda, Measuring sustainability: A multi-criterion framework, *Environment, Development and Sustainability* 7 (1) (2005) 117–134.
5. G. Munda, M. Nardo, On the methodological foundations of composite indicators used for ranking countries, *Ispra, Italy: Joint Research Centre of the European Communities* (2003) 1–19.
6. P. Lombardi, S. Giordano, H. Farouh, Y. Wael, An analytic network model for smart cities, in: *Proceedings of the 11th International Symposium on the AHP*, Sorrento, Naples, Italy, 15–18 June 2011, available: www.isahp.org.
7. P. Lombardi, S. Giordano, H. Farouh, W. Yousef, Modelling the smart city performance, *Innovation: The European Journal of Social Science Research* 25 (2) (2012) 137–149.

8. K. Shields, H. Langer, J. Watson, K. Stelzner, *European Green City Index: Assessing the Environmental Impact of Europes Major Cities*, Siemens AG: Munich, Germany, 2009.
9. G. Venkatesh, A critique of the European green city index, *Journal of Environmental Planning and Management* 57 (3) (2014) 317–328.
10. M. Nardo, M. Saisana, A. Saltelli, S. Tarantola, A. Hoffman, E. Giovannini, *Handbook on Constructing Composite Indicators*, OECD Publishing, 2005.
11. J. C. de Borda, Mémoire sur les élections au scrutin, *Histoire de l'Academie Royale des Sciences pour 1781* (Paris, 1784).
12. I. McLean, The Borda and Condorcet principles: Three medieval applications, *Social Choice and Welfare* 7 (2) (1990) 99–108.
13. J. G. Kemeny, L. Snell, Preference ranking: An axiomatic approach, in: J. G. Kemeny, J. L. Snell, *Mathematical Models in the Social Sciences*, MIT Press Edition, 1972, pp. 9–23.
14. C. Dwork, R. Kumar, M. Naor, D. Sivakumar, Rank aggregation methods for the web, in: *Proceedings of the 10th International Conference on World Wide Web*, ACM, 2001.
15. W. D. Cook, L. M. Seiford, Priority ranking and consensus formation, *Management Science* 24 (16) (1978) 1721–1732.
16. J. González-Pachón, C. Romero, Aggregation of partial ordinal rankings: An interval goal programming approach, *Computers Operations Research* 28 (8) (2001) 827–834.
17. M. E. Renda, U. Straccia, Web metasearch: rank vs. score based rank aggregation methods, in: *Proceedings of the 2003 ACM Symposium on Applied Computing*, ACM, 2003.
18. M. Farah, D. Vanderpooten, An outranking approach for rank aggregation in information retrieval, in: *Proceedings of the 30th Annual International ACM SIGIR Conference on Research and Development in Information Retrieval*, ACM, 2007.
19. M. S. Desarkar, S. Sarkar, P. Mitra, Preference relations based unsupervised rank aggregation for metasearch, *Expert Systems with Applications* 49 (2016) 86–98.
20. A. N. Langville, C. D. Meyer, *Who's# 1?: The Science of Rating and Ranking*, Princeton University Press, 2012.
21. E. Dopazo, M. L. Martnez-Céspedes, Rank aggregation methods dealing with incomplete information applied to smart cities, in: Fuzzy Systems (FUZZ-IEEE), *2015 IEEE International Conference on*, IEEE, 2015.
22. Z.-P. Fan, Y. Liu, An approach to solve group-decision-making problems with ordinal interval numbers, *IEEE Transactions on Systems, Man, and Cybernetics, Part B* (Cybernetics) 40 (5) (2010) 1413–1423.
23. E. Dopazo, M. L. Martnez-Céspedes, Rank aggregation methods dealing with ordinal uncertain preferences, *Expert Systems with Applications* 78 (2017) 103–109.
24. R. Ureña, F. Chiclana, J. A. Morente-Molinera, E. Herrera-Viedma, Managing incomplete preference relations in decision making: A review and future trends, *Information Sciences* 302 (2015) 14–32.

25. P. Berrone, J. E. Ricart, Iese cities in motion index 2016, available: http://www.iese.edu/research/pdfs/ST-0396-E.pdf (2016).
26. B. Roy, *Multicriteria Methodology for Decision Aiding*, Vol. 12, Springer Science & Business Media, Berlin, Germany, 2013.
27. T. Tanino, Fuzzy preference orderings in group decision making, *Fuzzy Sets and Systems* 12 (2) (1984) 117–131.
28. S. Brin, L. Page, Reprint of: The anatomy of a large-scale hypertextual web search engine, *Computer Networks* 56 (18) (2012) 3825–3833.
29. T. L. Saaty, How to make a decision: The analytic hierarchy process, *European Journal of Operational Research* 48 (1) (1990) 9–26.
30. E. Dopazo, J. González-Pachón, Consistency-driven approximation of a pairwise comparison matrix, *Kybernetika* 39 (5) (2003) 561–568.
31. E. Dopazo, M. Ruiz-Tagle, A parametric GP model dealing with incomplete information for group decision-making, *Applied Mathematics and Computation* 218 (2) (2011) 514–519.
32. E. Herrera-Viedma, F. Herrera, F. Chiclana, M. Luque, Some issues on consistency of fuzzy preference relations, *European Journal of Operational Research* 154 (1) (2004) 98–109.
33. G. A. Watson, *Approximation Theory and Numerical Methods*, John Wiley & Sons, 1980.
34. J. P. Keener, The Perron-Frobenius theorem and the ranking of football teams, *SIAM Review* 35 (1) (1993) 80–93.
35. C. W. Ueberhuber, *Numerical Computation 1: Methods, Software, and Analysis*, Springer Science & Business Media, 2012.
36. P. Nayak, S. Mishra, Efficiency of penas p2 distance in construction of human development indices, Technical report, University Library of Munich, Germany (2012).
37. N. Somarriba, B. Pena, Synthetic indicators of quality of life in Europe, *Social Indicators Research* 94 (1) (2009) 115–133.

CHAPTER 6

Agricultural Supply Chains Prioritization for Development of Affected Areas by the Colombian Conflict

Eduar Aguirre and Pablo Manyoma

CONTENTS

6.1 INTRODUCTION

Colombia has been immersed in an armed conflict for more than six decades, leaving fatal consequences for the country's development, especially in rural areas.

In general, the department of Cauca has been recognized as a strategic space for armed actors because of the convergence of communication channels between the center of the country and the Pacific Ocean (Figure 6.1). These characteristics favored the presence of insurgent groups and self-defense organizations, with a high incidence in the urban and especially rural population, through the presence of illegal armed groups and illicit crops.

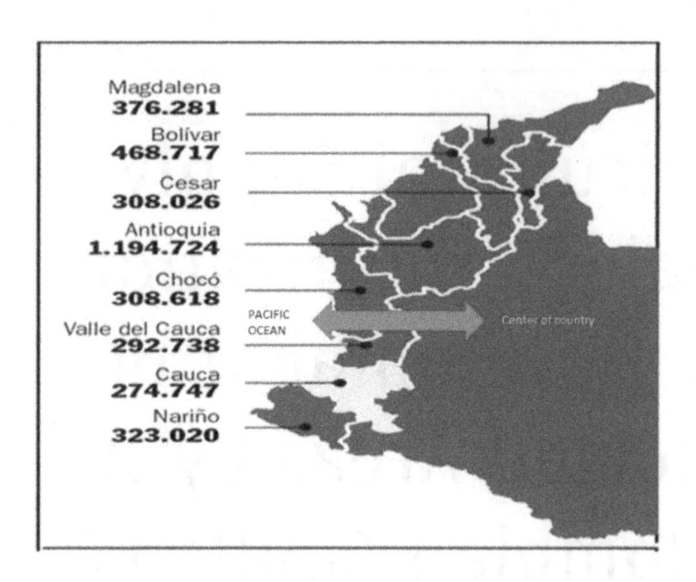

FIGURE 6.1 Communication channels between the center of the country and the Pacific Ocean. (From Own presentation by the authors based on *Semana* [Journal].)

Different problems have been identified, as well as some solutions proposed through the time. Nowadays, the Colombian government has been working to reestablish state control in most affected areas by the conflict through a strategic approach that integrates security, fight against drug trafficking, and economic and social development (USAID, 2016).

For this research, Cauca department's northern zone is our study object because it has become one of the most affected by this armed conflict (DNP, 2015). This zone is an area of great importance with approximately 400,000 inhabitants distributed in 13 municipalities. It is located in a geostrategic place; it has abundant natural resources, a great cultural richness represented in the ethnic diversity, peasant communities, religious traditions, urban zones, and low human development index (USAID, 2016).

In this subregion the highest economic activity of the department is concentrated; the most representative sectors of the economy of the subregion are industry, commerce, and agriculture. In the territory, five productive subsystems can be identified: the industrial-business in the flat zone; the commercial and services; the rural economy (mostly located in the mountain area); mining and environmental services. (Oficina internacional de los derechos humanos acción Colombia, 2016).

In contrast to these, the territory has high levels of poverty, inequality, and social conflict, in addition to current production models are not sustainable, in contradiction with the vocation of the soil and have been displacing the traditional production model.

The subregion has more than 356,164 hectares of arable land in different thermal floors, which allows planting of different products throughout the year, within a radius less than 60 km away from a large metropolitan area (Cali). Among identified products that can be converted into production chains are Cassava, Mango, Coffee, Pineapple, Gulupa, Avocado, Cocoa, Banana, Orange, Mandarin, and Lulo, among others (Figure 6.2).

To restore control in the most affected areas by the conflict, the Colombian government has been working through various institutions, such as Ministry of Agriculture and Rural Development, and Socio-Entrepreneurial Strengthening for Competitiveness, promoting the agro-industrial development of small producers to face the new agro-industrial markets.

Likewise, there are international initiatives such as the prioritization of municipalities by the United Nations, where it considers that nine of these municipalities are key to working in post-conflict Colombia (UNIVERSIDAD DEL VALLE, 2016). In addition, the US government,

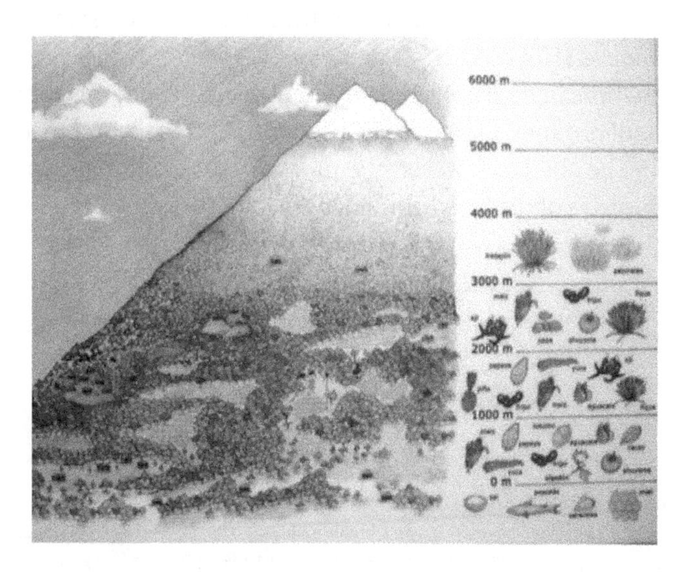

FIGURE 6.2 Thermal floors the northeastern subregion. (From PEDENORCA, Dirección nacional de planeación. Plan estratégico para el desarrollo del Norte del Cauca 2032. Contrato plan del Norte del Cauca, 2015, https://colaboracion.dnp.gov.co/CDT/_layouts/15/WopiFrame.aspx?sourcedoc=/CDT/Desarrollo%20Territorial/Archivo%20completo%20PEDENORCA%20NOV2015.pdf.)

through US Agency International Development (USAID), supports different efforts to promote economic prosperity, improve the living conditions of the most vulnerable populations, promote respect for human rights and justice, and address the natural threats generated by climate change (USAID, 2016).

The ability to coordinate efforts and resources becomes the central axis of new development model. Its trend is direct to the use of linkage schemes between producers and agribusinesses, which necessarily influences the evolution of agro-food chains and induces a greater vertical and horizontal coordination. This action implies to go from "market push" schemes to "market-pull" strategies aimed at meeting the needs of demand to increase the producers' capacity to adapt to continuous changes, according by Kaplinsky (Piñones et al., 2006).

In this perspective, productive agro-business approach is a support tool that allows actors from different agro-commercial chains in developing countries to insert or expand their participation in the market in a sustainable and competitive way. The concept of productive agro-business refers to the set of actors involved in the whole process of production, transformation, marketing, and distribution of similar goods. The stages and activities presents an agro-chain, developed in an environment of institutional and private services that directly influence its operation and competitiveness (Hoyos 2006) (Figure 6.3).

Agro-chains can be classified in different ways, depending, for example, on the type of product, the degree of differentiation or the number of actors involved in the chain. However, the most important thing is to consider that the objective of the classification is to facilitate the understanding and

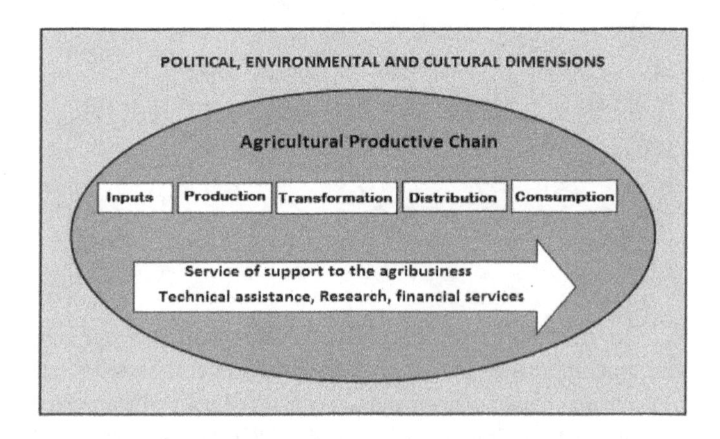

FIGURE 6.3 Basic structure of an agro-chain. (From CICDA 2004.)

analysis of the actors, links, and interrelationships that appear within the chain. The following are ways to classify agro-chains:

1. Depending on the type of product, its final use, and the degree of transformation or the characteristics of the demand, the agro-chains are classified as:
   - **Food chains**: those related only to fresh products.
   - **Industrial chains**: those related to products that receive some degree of transformation and nonfood products such as fibers, textiles, and leather.
2. Depending on the degree of differentiation of the product, the agro-chains are classified as:
   - Basic agro-chains: they revolve around products such as grains, tubers, and cereals. The basic agro-chains are characterized by a low demand elasticity, a low level of transformation, and a trade dominated by a small number of actors. Examples of this type of agro-chains are the rice, banana, wheat, and soybean chains.
   - Differentiated agro-chains: they are related to products that have special characteristics that differentiate them from commodities. This type of chains requires high coordination between producers, processors, and distributors. In addition, we can observe the existence of some degree of vertical integration between links. Examples of this type of chain are the chains of wine and organic products.
3. Depending on the type and number of actors involved, the agro-chains are classified as:
   - **Simple chains**: include only those actors and links directly related to the product in the different phases of production and marketing.
   - **Extended chains**: include, in addition to the main chain, other chains that may touch or interlock with it at some point, and which generally provide important inputs for obtaining the final product.

6.2 METHODOLOGY

Productive alliances are based precisely on a productive chain approach (including agro-chains), which is why their importance lies in the fact that they allow not only the market integration of all the links in the chain, including small producers (link primary), but also face the problems of information, financing, and technological innovation among productive agents and local public and private institutions that work around a given chain (Naciones unidas Colombia, 2014).

Decision problems are increasingly complex and usually involve several criteria. Precisely, the main force of the multi-criteria analysis is its ability to treat issues that have different situations in conflict, allowing an integrated evaluation of problem in question (Barfod 2012).

In this research, AHP is used to weight decision criteria and obtain their relative weight because the basis of the process rests on the fact that it allows numerical values of the judgments given by people, managing to measure how each element of the hierarchy contributes to the immediately superior level from which it emerges (Ávila 2000).

Then, Technique for Order of Preference by Similarity to Ideal Solution (TOPSIS) is used to compare a set of alternatives through the identification of weights for each dimension, the normalization of the scores of each dimension and the calculation of the distance between each alternative compared with the positive ideal alternative (the best in each dimension) and the ideal negative alternative (the worst), through the weighted dimensions using one of several possible distance measures (e.g., the Euclidean distance) and generating a ranking of alternatives according to their separation from positive and negative ideals (Huang et al. 2011). This work develops a ranking of agro-alimentary chains that may represent the geographical area previously defined.

6.3 RESULTS

The technical, economic, social, and environmental dimensions are most used in this type of decisions. Table 6.1 shows the relationship of the dimension with a selected criterion through the measurement of the performance of that criterion.

TABLE 6.1 Dimensions and Criteria Proposed

Dimension	Selected Criterion	Measurement Unit
Technical	Output	ton/ha
Economic	Price	US$/kg
Social	Affiliations	Number/agro-chains
Environmental	Impact	Expert opinion

Source: Own presentation by the authors.

- **Output (C1):** Shows relationship between production and harvested area. It must be maximized.
- **Price (C2):** Represents reference price of the goods produced by agro-chain. It must be maximized.
- **Affiliations (C3):** Organizations formally identified and operating around the product. It must be maximized.
- **Environmental impact (C4):** It is rated from 1 to 5, depending on the negative impact. It must be minimized.

Table 6.2 records the importance that one criterion has for the decision-making group over another, for this, the traditional scale (1–9) proposed by the AHP method.

In Table 6.3, we can observe the vector (magnitude) identified with the AHP. The criteria of the economic and social dimensions have similar relative importance, around 36% and 39%, respectively. Getting a RC = 9.8%

Through different reports and questions made to experts (Cauca Chamber of Commerce, USAID, Agronet, Asohofrucol, among others), it was possible to establish that the most promising products for the establishment of a agrichain are: Lulo, Cocoa, Plantain, Pineapple, and Avocado.

These products were selected as alternatives, in the following way:

A1: Lulo
A2: Cocoa
A3: Plantain

TABLE 6.2 Decision Matrix. Comparison of Criteria with Each Other

	C1	C2	C3	C4
C1	1	1/3	1/3	5
C2	3	1	1	3
C3	3	1	1	5
C4	1/5	1/3	1/5	1

Source: Own presentation by the authors.

TABLE 6.3 Relative Importance of the Criteria

	C1	C2	C3	C4
Relative Importance	18.53%	35.49%	38.84%	7.14%

Source: Own presentation by the authors.

A4: Pineapple
A5: Avocado

To develop the TOPSIS method, the following steps are proposed:

Step 1: The complete decision matrix (Table 6.4) shows each value (X_{ij}) that associates the alternatives or possible agrichains (A) with the criteria used; these values were determined according to the established form for each criterion.

Step 2: There is a need to normalize values because it is not the same as alternative 1 obtain a 5 in criterion 1 (C1) and a 5 in criterion 3 (C3). In Table 6.5, you can see all the values obtained and the differences when normalizing.

Step 3: To construct the weighted normalized decision matrix, the influence values of each weight were obtained in each criterion and in each alternative. These can be seen in Table 6.6.

TABLE 6.4 Decision Matrix. Result of Comparison of Criteria and Alternatives

	C1	C2	C3	C4
W	**18.53%**	**35.49%**	**38.84%**	**7.14%**
A1	10.00	0.62	6.00	3.00
A2	0.50	1.44	4.00	5.00
A3	7.00	0.31	6.00	3.00
A4	45.00	0.30	4.00	5.00
A5	8.00	0.64	3.00	3.00

Source: Own presentation by the authors.

TABLE 6.5 Normalized Matrix

		C1	C2	C3	C4
	W	**18.53%**	**35.49%**	**38.84%**	**7.14%**
n1	**A1**	0.21	0.35	0.56	0.34
n2	**A2**	0.01	0.82	0.38	0.57
n3	**A3**	0.15	0.18	0.56	0.34
n4	**A4**	0.95	0.17	0.38	0.57
n5	**A5**	0.17	0.37	0.28	0.34

Source: Own presentation by the authors.

TABLE 6.6 Normalized Weighted Matrix

		C1	C2	C3	C4
	W	**18.53%**	**35.49%**	**38.84%**	**7.14%**
n1	**A1**	0.039	0.126	0.219	0.024
n2	**A2**	0.002	0.292	0.146	0.041
n3	**A3**	0.027	0.063	0.219	0.024
n4	**A4**	0.176	0.061	0.146	0.041
n5	**A5**	0.031	0.130	0.110	0.024

Source: Own presentation by the authors.

Step 4: Determine the ideal positive solution and the ideal negative solution is assumed as a positive ideal solution the maximum value and as an ideal solution negative the minimum value that the variable X_{ij} has taken (Table 6.7).

Step 5: Calculation of distance measurements. The distance of each alternative to the ideal positive solution and to the ideal negative solution can be seen in Table 6.8.

Step 6: Calculation of proximity relative to the ideal solution and ordering of preferences; we obtain the relative proximity and ranking of possible alternatives (see Table 6.9).

TABLE 6.7 Positive and Negative Solutions for Each Criteria

	Max			**Min**
	C1	**C2**	**C3**	**C4**
A+	0.176	0.292	0.219	0.041
A−	0.002	0.061	0.110	0.024

Source: Own presentation by the authors.

TABLE 6.8 Measurements of Distance to the Ideal Solution

d+	A1	0.216	d−	A1	0.133
	A2	0.189		A2	0.235
	A3	0.274		A3	0.113
	A4	0.243		A4	0.179
	A5	0.244		A5	0.075

Source: Own presentation by the authors.

TABLE 6.9 Ranking Alternatives Evaluated

Ranking	Ri	Alternative
1	0.556	Cocoa
2	0.423	Pineapple
3	0.379	Lulo
4	0.291	Plantain
5	0.238	Avocado

Source: Own presentation by the authors.

Alternative 2 corresponds to the option of Cocoa as a viable product to start development of a basic agricultural supply chain. The main purpose of this work is to assign scarce public and international cooperation resources according with the needs of the region. This instance is a first approximation of work that must be done in near future to obtain that prioritization.

6.4 CONCLUSIONS

Alternative 2 corresponds to the option of Cocoa as a viable product to start with the formation of a basic agricultural supply chain.

At the moment the representative products of the region are: panela (derived from the sugar cane), coffee in different varieties, and some fruits of cold weather, such as strawberry, blackberry, among others.

The main purpose of this work is to assign scarce public and international cooperation resources according with the needs of the region.

REFERENCES

Agencia de Estados Unidos para el desarrollo internacional. Sobre Colombia. (2016, mayo). Available in https://www.usaid.gov/es/where-we-work/latin-american-and-caribbean/colombia.

Ávila, R.M. (2000). El AHP (Proceso Analítico Jerárquico) y su aplicación para determinar los usos de las tierras: El caso Brasil. Santiago de Chile: Proyecto regional "Información sobre tierras y aguas para un desarrollo agrícola sostenible." Cali, Colombia. Corporación autónoma regional del Valle del Cauca CVC.

Barfod, M.B. An MCDA approach for the selection of bike projects based on structuring and appraising activities. *European Journal of Operational Research*, 2012, vol. 218, no. 3, pp. 810–818.

Dirección nacional de planeación. (2015). Plan estratégico para el desarrollo del Norte del Cauca 2032. Contrato plan del Norte del Cauca. Available in https://colaboracion.dnp.gov.co/CDT/_layouts/15/WopiFrame.aspx?sourcedoc=/CDT/Desarrollo%20Territorial/Archivo%20completo%20PEDENORCA%20NOV2015.pdf.

Hoyos, Reinaldo. (27 de enero de 2016). Cauca, con 20 municipios priorizados para el posconflicto. El Nuevo Liberal. Disponible en http://elnuevoliberal.com/cauca-con-20-municipios-priorizados-para-el-posconflicto/.

Huang, I.B., Keisler, J. y Linkov, I. Multi-criteria decision analysis in environmental sciences: Ten years of applications and trends. *Science of the Total Environment*, 2011, vol. 409, no. 19, pp. 3578–3594.

Naciones unidas Colombia. (2014). Construcción de una paz territorial estable, duradera y sostenible. Sistema de la Naciones Unidas con apoyo de la Cooperación Alemana. Disponible enhttp://www.co.undp.org/content/dam/colombia/docs/MedioAmbiente/undp-co-pazyambiente-2015.pdf.

Oficina internacional de los derechos humanos acción Colombia. (2016, mayo). Extractive industries, natural resources and human rights in Colombia. Disponible en http://www.oidhaco.org/uploaded/content/ article/952080256.pdf.

Piñones, S., Acosta, L., y Tartanac, F. (2006). Alianzas Productivas en Agrocadenas Experiencias de la FAO en América Latina. Santiago, Chile. Organización de las Naciones Unidas para la Alimentación y la Agricultura.

UNIVERSIDAD DEL VALLE. Observatorio departamental Cauca (2016, mayo). Disponible en http://prevencionviolencia.univalle.edu.co/observatorios/cauca/departamental/archivos/perfil_cauca.pdf.

CHAPTER 7

Decision Making and Robust Optimization for Medicines Shortages in Pharmaceutical Supply Chains

João Luís de Miranda, Mariana Nagy, and Miguel Casquilho

CONTENTS

7.1 INTRODUCTION

The all set of Pharmaceutical Supply Chains (*PharmSC*) activities and operations are being covered by the fast-evolving information and communication technology (ICT) tools, such as cloud-based tools, Internet of Things (IoT) and Big Data analytics (BDA), and all the transportation of parts, jobs scheduling, and medicines distribution are being addressed by innovative approaches, as indicated by several researchers (e.g., Barbosa-Póvoa and Miranda, 2015; Barbosa-Póvoa et. al., 2016).

Finance and information fluxes are also focused when updating the *PharmSC* techniques, as well as the related decision making methodologies. Researchers from different European Union (EU) countries are addressing the updated definitions and goals on drugs shortages, adjusting the *PharmSC* approaches and models in a way to enlighten the costing and performance indexes of shortages. In the international cooperation scope, all the referred topics are being studied on behalf of the COST Action on *European Medicines Shortages Research Network—addressing supply problems to patients*, "Medicines Shortages" COST Action (CA15105).

The understanding of the entire *PharmSC* is often incomplete due to the large and complex set of inter-relations in the pharmaceutical networks and also because of the daily advancements of ICT tools. This text promotes the adequate methodologies to make full utilization of the available knowledge and data, namely when contradictory criteria arise. The multi-criteria decision-making (MCDM) methodologies need to be specifically aligned with the *PharmSC* enhancements in a useful manner to integrate robust modeling and to effectively drive impact on the pharmaceutical challenges.

The MCDM techniques reflecting the *PharmSC* challenges shall be implemented with the complete set of time and resource constraints. This issue is quite challenging because of the size and complexity of robust models, which typically address binary decisions (e.g., "Yes or No?"), nonlinear processes (e.g., design rules, production fluxes), uncertainties (products demands, energy costs, materials availabilities), and the interactions of *PharmSC* actors (suppliers, producers, wholesalers, retailers, and consumers) in a simultaneous way.

Thereafter, instead of the large-scale appreciation and integration of all the *PharmSC* aspects, many of the research works appear to concentrate on specific issues. Usually, each researcher has his own views and objectives, and developing a multidisciplinary and transdisciplinary framework is not common. In this way, many robust tools and models intended for *PharmSC* are not fully implemented because they are not properly programmed or tested. In practice, robust models hardly are used in *PharmSC* design because of their lack of applicability; robust models are often developed in a specific domain, and the real-world conditions are not fully considered.

Moreover, the necessity to integrate economic and financial criteria in the early design phases is becoming an imperative (Miranda, 2017). A new generation of medicines is emerging everyday from the pharmaceutical laboratories, stochastic and robust methods are still evolving, and the proper integration of MCDM tools will be a large benefit for the *PharmSC* managers and decision makers.

Within the *PharmSC* "lean bundles," two main principles are targeted: to accurately address medicines demand and to reduce wastes. The strategic and operational approaches designed to achieve a lean status for the complete *PharmSC* are providing better information for decision making, for improving financial and administrative activities, and by promoting the value created for the patients.

Appropriate MCDM approaches will contribute to better exploit *PharmSC* tools, data, and results, providing means to better describe the design process, at now only partially specified. That is, with the high level of data uncertainty, pharmaceutical innovation will benefit from the delivery of MCDM-based tools and from a deeper understanding of the pharmaceutical processes. Based on robust optimization (RO) models, the fields in the borders of pharmaceutical production can be also incorporated, such as environmental issues or managerial economics.

The purpose of this chapter is to present adequate MCDM tools and ROs to the medicines shortages within the *PharmSC* framework: in Section 7.2, the robust modeling for the design and scheduling of batch processes is revisited, so is the *PharmSC*'s problem structuring; in Section 7.3, the multiperiod robust model and the economic and financial estimators for medicines shortages are discussed; in Section 7.4, the results of the four MCDM tools are analyzed and compared; and finally, the main conclusions and future developments are presented in Section 7.5.

7.2 MEDICINES SHORTAGES AND THE ROBUST MODEL FOR DESIGN AND SCHEDULING *PharmSC* PROCESSES

In this section, we are introducing the COST Action on "Medicines Shortages" (CA15105), describing the main factors affecting the *PharmSC* operations with specific concern on manufacturing disruptions.

In addition, we use RO models for the design and scheduling of *PharmSC* processes, dealing with uncertainty and coping with computational issues.

By complementing the prior topics and also integrating MCDM subjects, the main purposes for the Action "Medicines Shortages" are:

1. The action-enhanced concepts and the qualitative methodologies for the drug shortages research;
2. The medicines manufacturing disruptions;
3. The provisions and procurement disruptions;
4. The clinic-pharmacological needs; and
5. The drugs shortages impact, both on patients and on healthcare systems.

The generalized model ***RObatch_ms***, as described by Miranda and Casquilho (2011, 2016), addresses the design of batch chemical processes and simultaneously considers the scheduling of operations. From a deterministic MILP model (Voudouris and Grossman, 1992), a model generalization is proposed to a stochastic 2SSP approach, this generalization being based on computational complexity studies (Miranda, 2011a, 2019). The generalized model for design and scheduling of batch chemical processes treats different time ranges, namely, the investment and scheduling horizons. Furthermore, the 2SSP framework allows the promotion of robustness solution, by penalizing the deviations; and the robustness in the model, with relaxation of the integrality constraints in the second-phase variables.

To better select the equipment to purchase, the optimal production policy must also be found because it directly affects the equipment sizing. However, it involves the detailed resolution of scheduling subproblems where decomposition schemes are pertinent. These subproblems are focused in the second phase of 2SSP, where the control variables (recourse) occur. The integer and binary variables related to the scheduling and precedence constraints are disregarded as control variables because they would make the treatment of the recourse problem difficult. Consequently, the second-phase variables are assumed continuous (e.g., the number of batches) and binary variables occur only in the first phase.

The study of existing models in the literature induces the enlargement of models and related applications (Miranda, 2007; Miranda and

Nagy, 2011), and this generalization of models simultaneously increases complexity and solution difficulties. A design and scheduling MILP model (Voudouris and Grossman, 1992) that seems to have no improvements for more than a decade is addressed. Analytical results and computational complexity techniques were applied to the referred deterministic and single-time MILP model, which is featuring multiple machines per stage, *zero-wait* (ZW) and *single-product campaign* (SPC) policy. That model was selected (Miranda, 2011b) because:

- For industrial applications with realistic product demands, multiple processes in parallel at each stage shall be considered, else numerical unfeasibility will occur;
- Due to modeling insufficiency, the option for the SPC policy arises from the difficulties to apply *multiple product campaign* (MPC) in a *multiple machine* environment;
- Assuming SPC, the investment cost will exceed in near 5% the cost of the more efficient MPC policy; then, the SPC sizing is a priori overdesigned, and this will permit to introduce new products or to accommodate unforeseen growth on demands.

The generalized model ***RObatch_ms***-adapted from Miranda and Casquilho (2016) along with some examples includes the optimization of long-term investment and also considers the short-term scheduling of batch processes. Indeed, deterministic models do not conveniently address the risk of a wider planning horizon, and scheduling decision models often deal with certain data in a single time horizon. Thus, difficulty increases when the combinatorial scheduling problem is integrated with the uncertainty of the design problem.

7.3 USEFULNESS OF THE MULTIPERIOD APPROACH

In terms of continuous variables and constraints, the dimension of ***RObatch_ms*** multiperiod model increases approximately linearly both with the number of periods, NT, and with the number of scenarios, NR (Miranda and Casquilho, 2016). Notwithstanding:

- The average solution time reveals perfectly exponential with reasonable times for the smaller instances, but it was no longer possible to obtain the optimum solution for larger instances. Some of the largest instances were also tested by enumeration, without practical results because the sparse character of the model, which presents a significant slice of coefficients and variables with null value.

- Also, a decision is taken to use a specific number of discrete scenarios ($NR = 5$) for the different instances. This decision is based on the comparison between two instances (Miranda and Casquilho, 2016): one with 5 scenarios; another with 10. And for the range of the penalization parameters tested, the configuration of the batch processes system kept unaltered despite the alteration in the number of scenarios.

Because the computational execution could not achieve an optimal exact solution, except for $\lambda qns = 0$, then estimators related to the best integer solution are reported in Table 7.1. Although optimal integer solution is not assured for each instance, the evolution pattern for technical estimators is related to the progressive increase of the selected dimensions in Table 7.2.

- The **robust NPV** is the objective function for the robust model; it is aiming to maximize the expected net present value (NPV) with robustness promoted by weighting and penalizing the expected value for the solutions variability, *dvtn*; the expected non-satisfied demand, *Qns*; and the expected capacity slackness, *slk*:

$$[\max]\Phi = \sum_{r=1}^{NR} prob_r \xi_r - \lambda dvt \sum_{r=1}^{NR} prob_r \cdot dvtn_r - \lambda qns \sum_{r=1}^{NR} \frac{prob_r}{NC \cdot NT} \left(\sum_{j=1}^{NC} \sum_{t=1}^{NT} Qns_{jtr} \right) - \lambda slk \sum_{r=1}^{NR} \frac{prob_r}{M \cdot NC \cdot NT} \left(\sum_{i=1}^{M} \sum_{j=1}^{NC} \sum_{t=1}^{NT} slk_{ijtr} \right) \tag{7.1}$$

- The **non-robust NPV** expected value, *Ecsi*, corresponds to the NPV probabilistic measure:

$$Ecsi = \sum_{r=1}^{NR} prob_r \xi_r \tag{7.2}$$

And the NPV at each scenario, r, considers the expected return minus the investment costs:

$$\xi_r = \sum_{j=1}^{NC} \sum_{t=1}^{NT} ret_{jtr} W_{jtr} - \sum_{i=1}^{M} \sum_{s=1}^{NS(i)} \sum_{p=1}^{NP(i)} c_{isp} y_{isp}, \ \forall r \tag{7.3}$$

7.3.1 The Technical Estimators

The technical estimators used in the appreciation of the best integer solution for multiperiod instances follow:

- (A) The **expected variability** corresponds to the negative deviation expected value:

$$Edvt = \sum_{r=1}^{NR} prob_r \cdot dvtn_r \tag{7.4}$$

Although the negative deviation is a linear probabilistic measure that penalizes only the NPV deviations below the NPV's expected value:

$$dvtn_r \geq \sum_{r'=1}^{NR} \left(prob_{r'} \xi_{r'}\right) - \xi_r \geq 0, \ \forall r \tag{7.5}$$

- (B) The **non-satisfied demand** expected value:

$$Ensd = \sum_{r=1}^{NR} \frac{prob_r}{NC \cdot NT} \left(\sum_{j=1}^{NC} \sum_{t=1}^{NT} Qns_{jtr} \right) \tag{7.6}$$

The non-satisfied demand, Qns, is the difference between the demanded quantities, Q_{jtr}, and the produced quantities, W_{jtr}; thus, it is defined as a slack variable through the couple of relations:

$$W_{jtr} + Qns_{jtr} = Q_{jtr}, \ \forall j,t,r \tag{7.7}$$

$$Qns_{jtr} \geq 0, \ \forall j,t,r \tag{7.8}$$

The percentage non-satisfied demand expected value is a relative measure, and the reference value is the expected demanded quantities, *Qmed*:

$$\%Ensd = \frac{Ensd}{Qmed} \cdot 100, \text{ with } Qmed = \sum_{r=1}^{NR} \frac{prob_r}{NC \cdot NT} \left(\sum_{j=1}^{NC} \sum_{t=1}^{NT} Q_{jtr} \right) \quad (7.9)$$

- (C) The **capacity slackness** is also defined as a probabilistic measure; it is the expected value for over-sizing in terms of the equipments volumes, *dv*, that are not fully used to produce the quantities, W_{jtr}:

$$Eslk = \sum_{r=1}^{NR} \frac{prob_r}{M \cdot NC \cdot NT} \sum_{j=1}^{NC} \sum_{t=1}^{NT} \left\{ \sum_{i=1}^{M} \sum_{s=1}^{NS} \sum_{p=1}^{NP} p(i).y_{isp}.\left(dv_{js} - S_{ij}.\frac{W_{jtr}}{n_{jtr}} \right) \right\} \quad (7.10)$$

The capacity slack in each instance is also defined as a slack variable through the relations set:

$$S_{ij} W_{jtr} + slk_{ijtr} = \sum_{s=1}^{NS(i)} \sum_{p=1}^{NP(i)} dv_{is} nc_{ijsptr}, \ \forall i, j, t, r \quad (7.11)$$

The percentage capacity slack expected value is a relative measure too, being the total volume of equipments defined as the reference value, *Vtotal*:

$$\%Eslk = \frac{Eslk}{Vtotal} \cdot 100, \text{ with } Vtotal = \sum_{i=1}^{M} \sum_{s=1}^{NS} \sum_{p=1}^{NP} \left(y_{isp} \cdot dv_{is} \right) \quad (7.12)$$

The alternatives on the design and sizing for a pharmaceutical batch plant with five stages (mixing, heating, reaction, separation, and drying) are defined by the order (*Ord*) of the discrete volume (*dv*) selected on a portfolio of six equipments/sizes (range 1–6) for each stage. For example:

- *Ord(s)* = (6/6/6/6/6) represents one single process or equipment by stage, with the larger sizes (#6) being selected for all these equipments and stages;
- *Ord(s)* = (11/11/11/11/11) represents the design alternative with two processes in parallel at each stage, with the lower sizes (#1) in every stage.
- *Ord(s)* = (111/666/222/555/444) represents a design alternative with three processes in parallel and with the same size or dimension at each stage, although different equipments, types, and sizes are allowed in the various stages, the same equipment and size is typically preferred for parallel configurations in the pharmaceutical industry because of operation, maintenance, quality assurance, and safety purposes.

The seven alternatives on design and sizing under analysis are presented in Table 7.1, along with the related penalty for non-satisfied demand, λqns, in the range [0–70]. The alternatives' set is thus representing the best configurations of process equipment for different service levels of the pharmaceutical plant, namely:

- Full service—the equipments and the sizes required to fully provide the drugs and medicines on demand are implemented (large penalization);

TABLE 7.1 The Seven Alternatives on the Design and Sizing and Service Levels within Pharmaceutical Plants

Alternatives	Penalty for Non-Satisfied Demand	Design and Sizing
	λqns	*Ord(s)*
	range	*range*
A_1	0	6/ 66/ 6/ 6/ 6
A_2	7	6/ 66/ 6/ 6/ 6
A_3	14	44/ 44/ 33/ 33/ 44
A_4	21	44/ 66/ 55/ 55/ 44
A_5	28	44/ 66/ 55/ 55/ 44
A_6	35	44/ 66/ 55/ 55/ 44
A_7	70	55/ 55/ 33/ 55/ 44

- Low service—the relaxation of most the demand requirements occurs, assuming the production orders could be not satisfied; the satisfaction of medicines demands are evaluated in face of their economic suitability (zero or very low penalization);
- Intermediate levels—intermediate levels for the relaxation of the demand requirements, assuming the hardness for the medicines demands is increasing with the penalization parameter (interim values).

Some redundancy may occur between the design alternatives A_1 and A_2 because the same batch design configuration is selected for the equipments and sizes; however the difference on the penalization parameter will drive different solutions for the medicines production that should be further detailed within the under the non-satisfied demand scope. Also, redundancy may occur between the three alternatives A_4, A_5, and A_6, but the associated design configuration, 44/ 66/ 55/ 55/ 44, presents a large spectrum of implementation that will deserve more detailed analysis pharmaceutical sector norms and regulations The related values are computed from the best integer solution of multiperiod instances, with $\lambda dvt = 1.0$, $\lambda slk = 0.1$, and with the λqns variation as described in Table 7.1 (Miranda and Casquilho, 2016).

Table 7.2 presents the values and weights for the technical estimators under analysis to better evaluate the batch design alternatives for different service levels of the pharmaceutical plant. By observation of Table 7.2, the initial strong decrease in the robust estimator *NPVrob*, is smoothing

TABLE 7.2 Technical Estimators for the Design and Sizing Alternatives

Criteria	Robust NPV	Expected NPV	Solutions Variability	Non-Satisfied Demand	Capacity Slackness
	NPVrob	***Ecsi***	***Edvt***	***%Ensd***	***%Eslk***
Alternatives	**Max**	**Max**	**Min**	**Min**	**Min**
A_1	217,727.51	259,301.90	21,806.00	11.4	7.0
A_2	98,543.73	259,085.06	21,934.99	11.2	6.9
A_3	103,766.40	177,575.18	21,430.64	1.6	6.0
A_4	93,458.55	148,858.35	21,403.17	0.1	7.6
A_5	92,085.91	148,858.35	21,403.17	0.1	7.6
A_6	90,713.27	148,858.35	21,403.17	0.1	7.6
A_7	98,040.86	141,819.05	21,135.75	0.0	6.2

with the corresponding decrease in non-satisfied demand; the non-robust estimator, *Ecsi*, also tends to decrease because the increasing return flows are not balancing the uprising costs; the variability estimator, *Edvt*, remains stable in the range from 21.1 to 21.9 ($\times 10^3$); the non-satisfied demand estimator, *Ensd*, is strongly reduced within the increase on discrete dimension levels, and then larger values of λqns are necessary to nullify this estimator; the capacity slack estimator, *Eslk*, remains stable in the range from 6.2% to 7.6%.

7.3.2 The Economic Estimators

The economic estimators used in the appreciation of the best integer solution for multiperiod instances follow too:

- (A) The **total cost** of equipments acquisition is assumed to occur only in the first period of the time horizon:

$$Ctotal = \sum_{i=1}^{M} \sum_{s=1}^{NS(i)} \sum_{p=1}^{NP(i)} c_{isp}\, y_{isp}, \ \forall r \tag{7.13}$$

- (B) The **benefit-cost ratio** on a percentage base is

$$\% Benef = 100 \frac{Ecsi + Ctotal}{Ctotal} \tag{7.14}$$

- (C) The ***payback* or *return of investment*** (ROI), in years, is obtained from the following relations set:

$$payback = t' - 1 + \frac{Ctotal - \sum_{t=1}^{t'-1} Ecash(t)}{Ecash(t')},$$

$$\text{with} \begin{cases} Ecash(t) = \sum_{j=1}^{NC} \sum_{r=1}^{NR} \left(prob_r \cdot ret_{jtr} \cdot W_{jtr} \right) \\ \exists^1 \ t' \ : \sum_{t=1}^{t'} Ecash(t) \geq Ctotal > \sum_{t=1}^{t'-1} Ecash(t) \end{cases} \tag{7.15a–c}$$

- (D) The **internal rate of return** (IRR) estimate is obtained through the discount return values to initial time period, *Ecash*0, and the estimate is taken as soon as,

$$\sum_{t=1}^{NT} Ecash0(t) \le Ctotal,$$

$$\text{with } Ecash0(t) = Ecash(t).\frac{(1+intrat)^t}{(1+IRR)^t} \qquad (7.16a\text{–}b)$$

Table 7.3 presents economic estimators associated to the best integer solution obtained in each of the multiperiod instances. The investment costs increase in direct relation with the progressive increasing of discrete dimension levels. Because of the satisfaction of almost all demand, there is short space of growth for return flows; thus, the successive increase of discrete dimensions and costs lead to a deterioration of the remaining economic estimators (*%Benef*, *payback*, IRR).

Although these economic estimators are based in the unitary return value of $ret_{jtr} = 0.15$, a virtual overestimation of production flows is promoted and corresponds to the ratio $\lambda qns/(NC.NT)$. Thus, for $\lambda qns = 7$, which corresponds to an overestimation of 0.2 on unitary return value, it more than doubles it, but the same dimensions configuration is selected as when $\lambda qns = 0$. In this manner, the robustness of

TABLE 7.3 Economic Estimators and Weights for the Design and Sizing Alternatives

Criteria	**Cost**	**Benefit-Cost ratio**	**Payback, Return of Investment**	**Internal Return Rate**
	Cost	***%Benef***	***payback***	**%IRR**
Alternatives	**Min**	**Max**	**Min**	**Max**
A_1	312,954.93	182.9	2.5	37.5
A_2	312,954.93	182.8	2.5	37.5
A_3	445,147.79	139.9	3.4	23.5
A_4	483,446.83	130.8	3.6	20.5
A_5	483,446.83	130.8	3.6	20.5
A_6	483,446.83	130.8	3.6	20.5
A_7	491,154.66	128.9	3.7	19.5

the reported solution is confirmed in the range of penalization parameters and in the context of an uncertain increase of unitary return values.

7.3.3 Additional Data

Additional data can be used to better detail the decision procedure, such as the characteristics within the attributes and the related indicators. Namely, the weighs for the criteria on service level, the economic estimators, and the technical estimators follow:

	Weighs
Service	0.652
Economic	0.217
Technical	0.130
sum	**1.000**

7.4 ANALYSIS OF RESULTS

In this section, four methods are used (the simple additive model; TOPSIS, ELECTRE, and the "e" Fuzzy model) and the related results are analyzed, and compared with the RO model. Both technical and economic estimators are studied; an enumerative ranking method and a joint analysis on technical and economic estimators are developed too. For that, consider the following notes:

- A global score for each pharmaceutical plant is computed by implementing the methods under analysis;
- The weights for attributes and characteristics are derived and analyzed in terms of discrepancy between final weights and individual judgments;
- The pharmaceutical plants are ranked by their score, and strengths and weaknesses are appreciated;
- A sensitivity analysis is performed by removing the QNS attributes (*Non-Satisfied Demands*) and changes in the final rank are checked; and
- An overall comment on the pharmaceutical alternative is provided, namely, addressing its strengths and its weaknesses, so as suggestions of improvements.

As there are seven alternatives and a number of technical and economic criteria (or attributes), a multi-criteria approach aims to build a top of the alternatives, selecting the one that best fits the criteria and sorting the possible action courses according to the concordance level to the criteria and weights. The decision matrix includes the consequences of each criterion on the considered alternatives.

Four of the most known methods are used for building the top of preferences:

1. The simple additive model—It is the most common direct MCDM that computes the general score of an alternative as a weighted sum of the consequences of the attributes (the utilities) with their weights.
2. The TOPSIS model—The *Technique for Order of Preference by Similarity to Ideal Solution* is an indirect MCDM that computes the score of each alternative based on its distance to the positive ideal solution and to the negative ideal solution.
3. The ELECTRE model—*The ELimination Et Choix Traduisant la REalité* is a family of complex MCDM that uses an algorithm based on the utility matrix and compares each pair of alternatives by their concordance and discordance.
4. The "e" Fuzzy model—It takes in consideration the fuzziness of the decisional situation and computes the score of each alternative by the degree of its membership to the fuzzy set of the ideal solutions.

For establishing the top of preferences given by each of the mentioned methods, dedicated software is used (Nagy and Miranda, 2013). The results are commented and the best solution is recommended on the basis of previous expertise and deployment on real cases. (Nagy and Negrușa, 2014).

7.4.1 Analysis on the Technical Estimators

The technical estimators on Table 7.2 are normalized through the Euclidian norm, that is, by square root of the sum of squares. The computed values for the normalized technical estimators and the related weighs are presented in Table 7.4.

The ranking of the alternatives, from the best to the weakest, obtained by applying the different decision methods based on the consequences of the technical criteria, is presented in Table 7.5.

TABLE 7.4 Normalized Matrix for Technical Estimators and Weights for the Design and Sizing Alternatives

Criteria	Robust NPV	Expected NPV	Solutions Variability	Non-Satisfied Demand	Capacity Slackness
	NPVrob	*Ecsi*	*Edvt*	*%Ensd*	*%Eslk*
Alternatives	**Max**	**Max**	**Min**	**Min**	**Min**
A_1	0.679	0.516	0.383	0.710	0.377
A_2	0.307	0.516	0.386	0.697	0.372
A_3	0.323	0.353	0.377	0.100	0.323
A_4	0.291	0.296	0.376	0.006	0.410
A_5	0.287	0.296	0.376	0.006	0.410
A_6	0.283	0.296	0.376	0.006	0.410
A_7	0.306	0.282	0.371	0.000	0.334
Weighs	**0.374**	**0.125**	**0.053**	**0.374**	**0.075**

TABLE 7.5 The Top of Preferences Considering the Technical Criteria

Additive Simple	Topsis	Electre	Fuzzy
A_1	A_7	A_7	A_7
A_7	A_4	A_3	A_1
A_3	A_5	A_1	A_2
A_4	A_6	A_2	A_3
A_5	A_3	A_4	A_4
A_6	A_1	A_5	A_5
A_2	A_2	A_6	A_6

The ranking is slightly different due to the different algorithms used by the different methods. The alternative A_7 is the preferred course of action if only the technical criteria are considered. The second-best alternative is not uniquely determined; good courses of action are A_3, A_1, and A_4, the choice depending on further evaluation.

7.4.2 Analysis on the Economic Estimators

The economic estimators on Table 7.3 are also normalized through the Euclidian norm, being the computed values and the related weighs presented in Table 7.6. To avoid duplication, the probabilistic measure on the expected NPV is not included here because this estimator was already addressed in the analysis of technical estimators, as described in the prior section.

TABLE 7.6 Economic Estimators and Weights for the Design and Sizing Alternatives

Criteria	Cost	Benefit-Cost ratio	Payback, Return of Investment	Internal Return Rate
	Cost	*%Benef*	*payback*	%IRR
Alternative	**Min**	**Max**	**Min**	**Max**
A_1	0.271	0.465	0.286	0.530
A_2	0.271	0.465	0.286	0.530
A_3	0.385	0.356	0.388	0.332
A_4	0.418	0.333	0.411	0.290
A_5	0.418	0.333	0.411	0.290
A_6	0.418	0.333	0.411	0.290
A_7	0.425	0.328	0.423	0.276
Weighs	**0.374**	**0.125**	**0.053**	**0.374**

The top of preferences built considering only the consequence of the economic criteria are presented in Table 7.7.

The rankings based on the economic criteria are similar for the different methods. The results are quite opposite to the previous ones: the best alternative is A_1 and the weakest is A_7. As expected, the best technical result bears with the highest economic costs. From the two tables, the optimal alternative seems to be A_1 or A_3. For a better view on the problem and its solutions, further considerations are made, namely the input given by the design and sizing of the pharmaceutical plants is optimized and only the best five alternatives are considered.

TABLE 7.7 The Top of Preferences Considering the Economical Criteria

Additive Simple	Topsis	Electre	Fuzzy
A_1	A_1	A_1	A_1
A_2	A_2	A_2	A_2
A_3	A_3	A_3	A_3
A_4	A_4	A_6	A_4, A_5, A_6
A_5	A_5	A_5	A_7
A_6	A_6	A_4	#
A_7	A_7	A_7	#

7.4.3 The Enumeration Search Ranking

An enumerative procedure of all the foreseen configurations is developed for $\lambda qns = 0$: by fixing binary variables, all the alternative configurations are examined without penalizing the non-satisfied demands. In this way, the robust model ***RObatch_ms*** is directed to better focus the economic estimators because the capacity slackness presented a narrow range of variation for all the design alternatives under analysis: the technical estimator *%Eslk* was about 6.2%–7.6% (Table 7.2). The complete examination of the solutions set ($7^5 = 16807$ configurations) also allowed the confirmation of the optimal solution for the robust model (Miranda and Casquilho, 2016) and the best five design alternatives (B1–B5) computed for $\lambda dvt = 1.0$ and $\lambda slk = 0.1$ are described in Table 7.8.

In this way, a sensitivity analysis is performed by removing the attributes and penalties for non-satisfied demands, while the alterations in design configurations are checked.

The economic estimators for these best five design configurations are ranked in descending order of non-robust estimator, *Ecsi*, and reported in Table 7.9. This enumeration of the best integer solutions allows insight of the configurations evolution within this economical estimator. Although the selection of configurations is performed by the non-robust value, *Ecsi*, instead of the robust value, *NPVrob*, these two estimators point at the same exact solution. Moreover, additional data is provided about the best alternative configurations, and this information is useful to build local search heuristics.

Then, although different policies for risk treatment can be applied when considering larger horizons or when highly uncertain environment

TABLE 7.8 The Five Ranked Alternatives on the Design and Sizing of Pharmaceutical Plants

Alternatives	Penalty for Non-Satisfied Demand	Design and Sizing
	λqns	*Ord(s)*
	(Min)	*(Min)*
B1	0.0	6/ 66/ 6/ 6/ 6
B2	0.0	6/ 55/ 6/ 6/ 6
B3	0.0	6/ 44/ 6/ 6/ 6
B4	0.0	6/ 66/ 6/ 6/ 22
B5	0.0	6/ 55/ 6/ 6/ 22

TABLE 7.9 Economic Estimators and Weighs for the Five Ranked Alternatives

Criteria	**Cost**	**Expected NPV**	**Benefit-Cost Ratio**	**Payback, Return of Investment**	**Internal Return Rate**
	Cost	***Ecsi***	***%Benef***	***payback***	**%IRR**
Alternatives	**Min**	**Max**	**Max**	**Min**	**Max**
B1	312,954.93	259,301.90	182.9	2.5	37.5
B2	309,896.72	250,556.72	180.9	2.5	36.5
B3	302,602.29	192,758.20	163.7	2.8	31.5
B4	326,965.58	144,114.30	144.1	3.2	24.5
B5	323,907.37	141,304.49	143.6	3.3	24.5

is considered (obsolescence, technologic innovation, short life cycle of products), the utilization of a limited number of time periods NT can be suitable. Considering complex contexts that require more detailed studies, a larger horizon or a larger number of periods is to be considered. If intending to compare alternative configurations, or aiming to confirm exactness of a given configuration, an enumerative procedure can be useful if the number of alternatives is limited.

The normalized estimators and the related weighs follow in Table 7.10.

The ranking of the alternatives given by the four methods is presented in Table 7.11.

TABLE 7.10 Normalized Matrix for the Economic Estimators and Related Weighs

Criteria	**Cost**	**Expected NPV**	**Benefit-Cost Ratio**	**Payback, Return of Investment**	**Internal Return Rate**
	Cost	***Ecsi***	***%Benef***	***payback***	**%IRR**
Alternative	**Min**	**Max**	**Max**	**Min**	**Max**
B1	0.444	0.569	0.499	0.388	0.534
B2	0.439	0.550	0.494	0.388	0.520
B3	0.429	0.423	0.447	0.435	0.449
B4	0.464	0.316	0.393	0.497	0.349
B5	0.459	0.310	0.392	0.512	0.349
Weighs	**0.125**	**0.374**	**0.053**	**0.374**	**0.075**

TABLE 7.11 The Top of Preferences Considering only the Economical Best Five Alternatives

Additive Simple	Topsis	Electre	Fuzzy
B1	B1	B1	B1
B2	B2	B2	B2
B3	B3	B3	B3
B4	B4	B4	B4
B5	B5	B5	B5

By selecting the best five configurations from the economic point of view, the results are consistent with the previous ones. The slightly different ranking with Electre is due to the influences of the score of each alternative on the others—a specificity of Electre. Moreover, it has to be specified that the scores are computed according to different algorithms and the results are very similar.

7.4.4 Joint Analysis on Technical and Economic Estimators

As previously shown, the technically best alternative comes with the highest financial costs, being the worse if considering only the economic criteria. A joint analysis is proposed, considering a set of technical and economic criteria, chosen as to be significant and eliminating any redundancy. The normalized matrix is presented in Table 7.12.

By applying the same four MCDM, the results in Table 7.13 are obtained.

The top of preferences is a "negotiation" between the two types of criteria—technical and economic estimators. Alternative A_1 is given as the best, followed by A_2. The following positions are divided by A_3, A_7, and A_4, these alternatives being acceptable solutions both from economic and technical point of view.

The results are closer to the top of preference built according to the economic criteria because the values are weighted by a human decision maker for whom the economic point of view seems to be more important. If penalizing the consequences for the non-satisfied conditions, the robust model is more accurate.

TABLE 7.12 Normalized Matrix for a Set of Joint Technical and the Economic Estimators and Related Weighs

Criteria	Cost	Benefit-Cost Ratio	Payback ROI	Internal Return Rate	Robust NPV	Expected NPV	Solutions Variability	Non-Satisfied Demand	Capacity Slackness
	Cost	*%Benef*	*Payback*	%IRR	*NPVrob*	*Ecsi*	*Edvt*	*%Ensd*	*%Eslk*
Alternatives	**Min**	**Max**	**Min**	**Max**	**Max**	**Max**	**Min**	**Min**	**Min**
A_1	0.271	0.465	0.286	0.530	0.679	0.516	0.383	0.710	0.377
A_2	0.271	0.465	0.286	0.530	0.307	0.516	0.386	0.697	0.372
A_3	0.385	0.356	0.388	0.332	0.323	0.353	0.377	0.100	0.323
A_4	0.418	0.333	0.411	0.290	0.291	0.296	0.376	0.006	0.410
A_5	0.418	0.333	0.411	0.290	0.287	0.296	0.376	0.006	0.410
A_6	0.418	0.333	0.411	0.290	0.283	0.296	0.376	0.006	0.410
A_7	0.425	0.328	0.423	0.276	0.306	0.282	0.371	0.000	0.334
W_j	0.267	0.038	0.267	0.053	0.140	0.047	0.020	0.140	0.028

TABLE 7.13 The Top of Preferences Considering the Joint Set of Technical and Economic Criteria

Additive Simple	Topsis	Electre	Fuzzy
A_1	A_1	A_1	A_1
A_2	A_2	A_2	A_2
A_3	A_3	A_3	A_7
A_4	A_4	A_7	A_3
A_5	A_5	A_4	A_4
A_6	A_6	A_5	A_5
A_7	A_7	A_6	A_6

7.5 CONCLUSIONS

The study of optimization cases and models from the literature allowed a detailed overview and permitted to conjugate realistic subjects both in formulation and solution procedures. While developing theoretical studies, the various models at hand are detailed and insight is gained, their benefits and limitations are balanced, and robust generalization is developed. In addition, the studies on computational complexity along with the computational implementation fostered the construction of heuristics, such as local search procedures.

Based on the generalization approach described in prior paragraph, the *Batch* problem is addressed:

- The treatment of uncertainty also considered the problems' specificities;
- The short-term scheduling and the multi-period horizon were simultaneously addressed;
- The deterministic approach from the literature is generalized onto a stochastic one, and economic parameters of interest were evaluated.

Further developments include modeling issues and solution methods, and the development of decision support systems will foster the application to industrial cases.

ACKNOWLEDGMENTS

Authors thank: (JLM) *Escola Superior de Tecnologia e Gestão* at *Instituto Politécnico de Portalegre* (ESTG/IPP) and CERENA-*Centro de Recursos Naturais e Ambiente* at Instituto Superior Técnico (IST); (MN) the

Department of Mathematics and Computer Science (UAV) and the UAV researchers who contributed to implementing and testing the software; and (MC) the Department of Chemical Engineering (IST), and CERENA. This work was partially developed at CERENA (strategic project FCT-UID/ECI/04028/2019).

REFERENCES

Barbosa-Póvoa, A.P., Corominas, A., Miranda, J.L., *Optimization and Decision Support Systems for Supply Chains* (Springer, Cham, 2016).

Barbosa-Póvoa, A.P., Miranda, J.L., *Operations Research and Big Data, Proceedings of IO2015-XVII Portuguese OR Conference* (Springer, Cham, 2015).

COST Action CA15105 European Medicines Shortages Research Network—Addressing supply problems to patients (Medicines Shortages) http://www.cost.eu/COST_Actions/ca/CA15105 (access in January 31, 2019).

Miranda, J.L., Robust Optimization and Technical-Economic Estimators for Pharmaceutical Supply Chains, EAHP2017—22nd *Congress of the European Association of Hospital Pharmacists*, special session in "COST Action CA15105-Breakthrough networking and sharing responsibilities to cope with the Medicines Shortages challenge" (Cannes, France, March 22–24, 2017).

Miranda, J.L. (2011a) Computational complexity studies in the design and scheduling of batch processes. In: Book of Abstracts of IO2011—15th Congress of APDIO, University of Coimbra, Coimbra, Portugal, April 18–20, 2011.

Miranda, J.L. (2011b) The design and scheduling of chemical batch processes: building heuristics and probabilistic analysis. *Theory and Applications of Mathematics & Computer Science*. 1(1): 45–62.

Miranda, J.L. (2019) The design and scheduling of chemical batch processes: Computational complexity studies, *Computers and Chemical Engineering* 121: 367–374. doi:10.1016/j.compchemeng.2018.11.011 (access in 31-January-2019).

Miranda, J.L., Casquilho, M., *Optimization Concepts:* II—*A More Advanced Level*, pages 79–97, In: *Optimization and DSS for SC*, edited by Ana Póvoa, Albert Corominas and João Miranda in "Lectures Notes on Logistics" series at Springer Verlag (Springer, Cham, 2016).

Miranda, J.L., Casquilho, M. (2011) Design and scheduling of chemical batch processes: Generalizing a deterministic to a stochastic model. *Theory and Applications of Mathematics & Computer Science*. 1(2): 71–88.

Miranda, J.L., Nagy, M. (2011) A case of cooperation in the European Operations Research education. *European Journal of Engineering Education* 36(6): 571–583.

Nagy, M., Miranda, J.L. (2013) Computer Application for Interactive Teaching of Decision Making Methods. *AWER Procedia Information Technology & Computer Science* 3: 1584–1589.

Nagy, M., Negruşa, A. (2014) Using Electre Method for a Computer Assisted Decision in the Field of Public Acquisition in Romania, in *Proceedings of Globalization and Intercultural Dialogue: Multidisciplinary Perspectives*, 194–199.

Voudouris, V.T., Grossmann, I.E. (1992) Mixed-Integer Linear Programming reformulations for batch processes design with discrete equipment sizes. *Industrial & Engineering Chemistry Research*, 31: 1315–1325.

CHAPTER 8

Using Spatial Decision Models for Rank Ordering Chocolates

Valérie Brison, Claudia Delbaere,
Koen Dewettinck, and Marc Pirlot

CONTENTS

8.1 INTRODUCTION

Many multiple criteria decision problems take place in a spatial context. Well-known ones are location problems. For example, we may want to find the best places for housing (Joerin, 1998) or to site a corridor for a road or a railway track (Chakhar, 2006, Aissi et al., 2012). In such problems, maps play an essential role. Indeed, the decision-aiding process usually leads to a map describing the potential of the territory for a given use at a given time. Such a map serves as a basis for the decision. Using or integrating decision-making tools methods in geographic information systems has been an active research direction in recent years (Malczewski, 2006, 2010, Sobrie et al., 2013).

Another type of spatial decision problems deals with maps representing the potential of the territory for a given use but at different times. Consider the case study developed by Metchebon (2010) (see also Metchebon et al., 2015). The authors were concerned with evaluating land degradation in a region of Northern Burkina Faso, specifically, the Loulouka catchment basin. A map representing the region under study is divided into regular squares of 25 hectares. Each square is assessed on several criteria that reflect the factors that have an impact on land degradation. The ELECTRE Tri (Roy and Bouyssou, 1993) method is applied to aggregate the information and to assign each square to one of the four predefined categories representing the response to the risk of degradation of the landscape, namely "Adequate," "Moderately Adequate," "Weakly Adequate," and "Not Adequate." This leads to a map representing the situation at a given time. Assume that a managerial policy is applied to improve the state of the region for instance by setting up an educational program aiming to promote sustainable agricultural methods. To assess the effectiveness of the policy, the process briefly described is reused on the new data obtained several years after the application of the policy has started. This leads to a second map representing the new situation. The two maps obtained are called *decision maps.*

Figure 8.1 represents the state of the region before the application of a specific policy aiming at improving the state of the region, and Figure 8.2 represents the state of the same region 5 years after its application. By comparing the new map with the old one, it can be observed that the state of some squares has improved whereas it has deteriorated for some others. The issue is to know whether the state of the region has globally improved or become worse. The objective is thus to develop models that will help a decision maker to express his or her preferences over maps like these represented in Figures 8.1 and 8.2.

The approach used consists in developing an axiomatic characterization of the decision-maker's preference guaranteeing that a given model can be used to represent his or her preference and in proposing an elicitation process to determine the model's parameters.

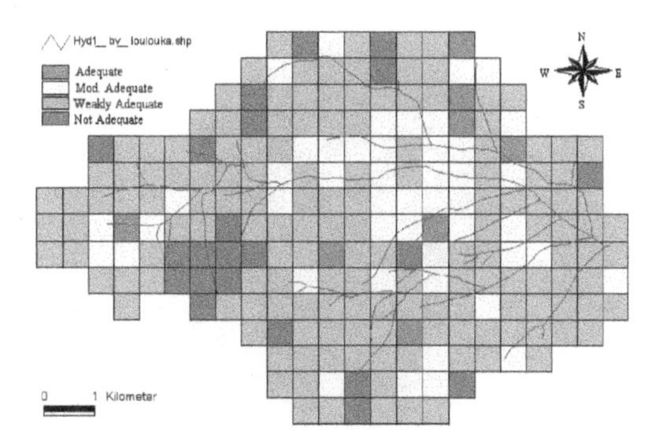

FIGURE 8.1 Previous state of the Loulouka basin.

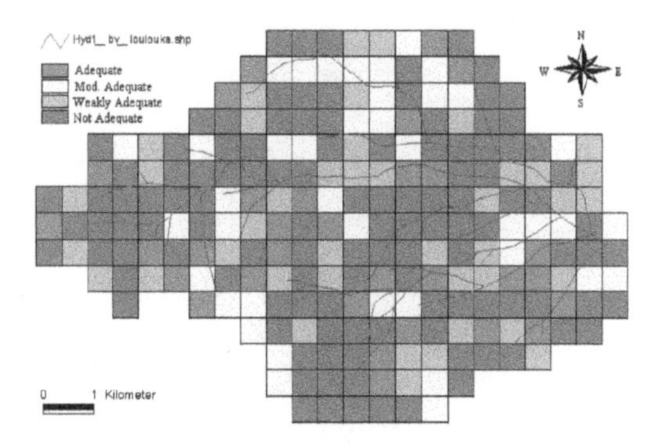

FIGURE 8.2 Current state of the Loulouka basin.

These models are not only relevant in a geographic context but more generally for comparing and ranking images. The present work deals with the comparison of photographs of (filled) chocolates during storage, an application that interests the chocolate industry. More precisely, this industry has to deal with the *fat bloom* phenomenon (i.e., the appearance of a white-grayish layer or white spots on the chocolate surface due to recrystallization of the fat). This phenomenon is more abundant in filled chocolates, the so-called pralines because oil migrating from the soft center to the chocolate shell accelerates the fat recrystallization and the formation of visible fat bloom. As illustration, Figure 8.3 shows a fresh praline that is not bloomed, while the surface of the chocolate in Figure 8.4 is completely covered with fat bloom. Experts are appointed to rank order chocolate samples according to the degree of fat bloom affecting them.

The rest of this chapter is organized as follows. In Section 8.2, we briefly describe some of the models we have developed to represent

FIGURE 8.3 Not bloomed praline.

FIGURE 8.4 Bloomed praline.

the decision-maker's preference over maps and, more generally, spatial objects. The application of these models to rank order chocolate images is presented in Section 8.3. Section 8.4 is devoted to some conclusions and perspectives.

8.2 REPRESENTING PREFERENCES OVER MAPS

This section is devoted to the presentation of two models we have developed to represent the decision-maker's preference over maps. Each model differs in the spatial information taken into account in the comparison. The first one deals with the simple case where no spatial aspect has an influence on the decision-maker's preference. The second one deals with the case in which such aspects do have an influence, in particular, the case in which contiguity is taken into account.

Throughout this chapter, maps are divided into polygons or pixels, each of them being assessed on the same scale consisting of a finite number of ordered categories. The decision-maker's preference over such spatial objects is a binary relation $\succsim$. It is supposed to be a complete pre-order (complete, reflexive, transitive relation). Its asymmetric part is denoted by $\succ$ and its symmetric part by $\sim$ (indifference relation).

8.2.1 Comparing Maps in a Basic Way

The main assumption here is that the region under study is homogeneous. This means that the only thing that matters to the decision maker is the area distribution in categories of the whole maps, that is to say the proportion of the surface area assigned to each category. If two maps have the same distribution in categories, they are supposed to be indifferent.

In this context, the model used to represent the decision-maker's preference is an expected utility model. This model consists in associating a utility value, u_i, with decision maps completely assigned to category i. From these utilities, the expected utility of any decision map can be computed, simply by weighting the former by the proportion of the surface actually assigned to category i. More formally, this model assigns value $u(A)$ to decision map A with

$$u(A) = \sum_{i=1}^{n} x_i(A)u_i, \tag{8.1}$$

where n is the number of categories and $x_i(A)$ is the proportion of the surface area of map A assigned to category i. Decision map A is preferred to decision map B if $u(A)$ is greater than $u(B)$. This model was axiomatized and an elicitation process has been designed to determine the model's parameter (i.e., the utilities, u_i) by interacting with the decision maker. For more details about this model, the reader is referred to Metchebon et al. (2013) or Brison (2017).

8.2.2 Taking Contiguity into Account

It can be the case that contiguity has an influence on the decision-maker's preference. Consider the situation illustrated in Figures 8.5 and 8.6. The two maps have the same number of pixels assigned to each category. The homogeneity of the region would imply that these maps are equivalent because the distributions of the pixels among the categories are the same. However, in some cases, the decision maker could prefer a map offering large contiguous areas in the same category, or, alternatively, the decision maker could favor a random distribution.

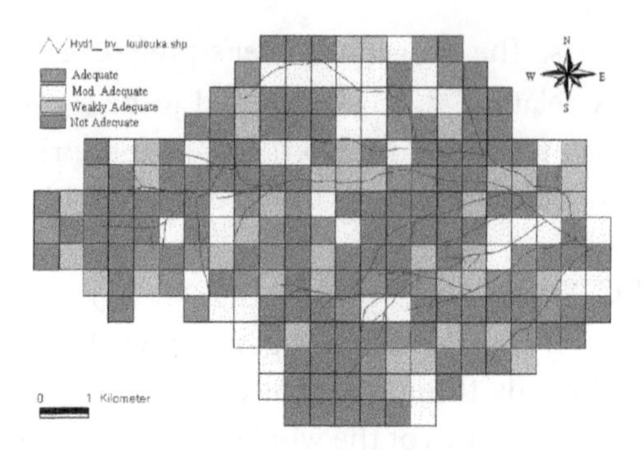

FIGURE 8.5 Scattered map.

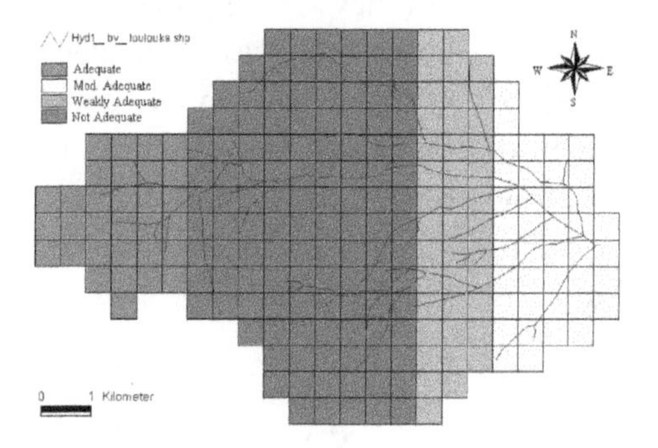

FIGURE 8.6 Clustered map.

To take this kind of aspects into account, a model based on the Choquet integral was proposed by Metchebon (2010) (see also Metchebon et al., 2013) and axiomatized by Brison (2017). Such a model is able to take "interactions" into account. In the context of maps comparison, two polygons or pixels interact if they are contiguous. Obviously, there are many ways to define the notion of contiguity. For example, two pixels could be considered as contiguous if they have at least one common border, or at least one common point. Even interactions with more distant pixels could be considered.

This model is briefly presented hereafter, without going into the theoretical details. Consider model (8.1) that can be used when no geographic aspect matters. This model can be written in the following way:

$$
\begin{aligned}
u(A) &= \sum_{i=1}^{n} x_i(A) u_i \\
&= \sum_{s \in S} m(\{s\}) u(s)
\end{aligned}
\tag{8.2}
$$

where S is the set of polygons or pixels s constituting the map, $m(\{s\})$ is the proportion of the surface area s in the whole map and $u(s)$ is equal to u_i if s is assigned to category i. For taking into account interactions between two pixels s, t, terms are added to model (8.2):

$$u(A)=\sum_{s\in S} m(\{s\})u(s)+\sum_{s,t\in S} m(\{s,t\})\min(u(s),u(t)).$$

For taking into account interactions between three pixels r, s, t, other terms are added:

$$u(A)=\sum_{s\in S} m(\{s\})u(s)+\sum_{s,t\in S} m(\{s,t\})\min(u(s),u(t))$$

$$+\sum_{r,s,t\in S} m(\{r,s,t\})\min(u(r),u(s),u(t)).$$

More generally, considering all possible interactions leads to

$$C_m(u,A)=\sum_{Y\subseteq S} m(Y)\min_{s\in Y} u(s). \tag{8.3}$$

Restrictions have to be imposed on m in order to guarantee that $C_m(u,A)$ is a Choquet integral (see Chateauneuf and Jaffray, 1989, Proposition 2, p. 265).

Up to this point, only maps comparison has been considered. What about comparing images? An image consists of a set of pixels that have an evaluation representing their color on a given scale. Basically, a map is an image that represents some geographic features. Since it is possible to compare maps divided into pixels that have an evaluation on a finite scale, it is also possible to compare images.

For example consider Figures 8.7 and 8.8 representing, for illustration purpose, two (fictitious) pralines that have the same distribution of white pixels; one shows a big white spot in the middle while several small white spots are scattered over the surface of the other. A consumer might strictly prefer one of these pralines over the other, although they have the same number of white pixels. A Choquet integral is able to discriminate between these pralines, but model (8.1) will consider them as indifferent. Indeed, because the distribution of the pixels on the gray scale is the same for both pralines, their expected utility will be the same. Also, consider the Choquet integral and let us define two pixels as contiguous if and only if they have a common border. In this case, we have

FIGURE 8.7 Praline with a big white spot.

FIGURE 8.8 Praline with scattered white spots.

$$C_m(u, A) = \sum_{s \in S} m(\{s\})u(s) + \sum_{\substack{s,t \in S \\ \{s,t\} \text{ contiguous}}} m(\{s,t\}) \min(u(s), u(t)). \quad (8.4)$$

Giving the same weight to each pixel, the first sum will be the same for both pralines; the only difference is in the second sum. Setting the same positive (resp. negative) value to $m(s,t)$ will give a better evaluation to the praline in Figure 8.7 (resp. Figure 8.8). Table 8.1 shows an example of values obtained for the second sum in model (8.4).

TABLE 8.1 Examples of Evaluation Obtained for Figures 8.7 and 8.8

	$\frac{1}{M}\sum_{\substack{s,t\in S\\ \{s,t\}\text{ contiguous}}}\min(u(s),u(t))$	$\frac{1}{M-4N}\sum_{\substack{s,t\in S\\ \{s,t\}\text{ contiguous}}}\min(u(s),u(t))$
Figure 8.7	193.93	−193.26
Figure 8.8	193.77	−193.09

Note: N is the total number of pixels and M is the total number of contiguous pairs of pixels.

8.3 COMPARING PICTURES OF BLOOMED CHOCOLATES

This section deals with pictures representing pralines. The issue is to compare these images with respect to the presence of *fat bloom*, a natural phenomenon that is typically related to chocolate products. Before explaining how the models described in Section 8.2 can be used, we briefly explain what fat bloom is and which data are available.

8.3.1 The Fat Bloom Phenomenon

When a chocolate bar, a praline, or even a chocolate covered biscuit is left for too long inside a cupboard or in a too warm place, a white-grayish layer or white spots can be observed on their surface. This phenomenon is called *fat bloom* and is caused by recrystallization of the fat, followed by growth of fat bloom crystals on the chocolate surface. Actually, fat bloom development is characterized by two aspects: the disappearance of gloss and the appearance of whitish areas.

The fat bloom phenomenon is the most important reason for shelf life limitation of chocolate products. Unfortunately, fat bloom cannot be avoided, even in optimal storage conditions. Consequently, studying fat bloom is an important issue for the chocolate industry. Indeed, it is necessary to understand where it comes from to be able to find solutions that delay its occurrence, as much as possible. Some studies have been led, not only to understand the mechanism of fat bloom (see, e.g., Reinke et al., 2015) but also to quantify fat bloom (see, e.g., Nopens et al., 2008). The present work relates to the latter trend.

8.3.2 Fat Bloom Quantification

Fat bloom quantification is usually performed by a trained panel of experts who give a score to chocolate products whose state regarding fat bloom evolves with time. In the study realized by Nopens et al. (2008), several sets of chocolates were used that differ by their composition. A panel of experts was asked to evaluate fat bloom by giving a score to the samples on a low-resolution scale (ordinal scale with six levels). The scoring was performed every 2 weeks. Of particular interest for us is the fact that, at a given time, several pieces of chocolate of the same type were assessed using the same scale. From these evaluations, a ranking of the samples, from the best (i.e., least bloomed) to the worst (i.e., most bloomed) is easily derived. The goal, in the present work, is to examine whether the experts' ranking can be retrieved from the analysis of the chocolate pictures by using one of the models we have developed. The experiments we present in the rest of this section were realized by "Cacaolab," a spin-off of Ghent University, and concerns the comparison of pralines.

In practice, it turned out that "contiguity" may have an influence, in the sense that a chocolate presenting several small scattered white spots is preferred to a chocolate presenting a large whitish area. Indeed, non-expert people like consumers may miss noticing very small white spots, whereas large whitish zones attract their attention. Consequently, an expert panel consisting of 7–9 people were asked to compare pralines on the basis of two questions.

Question 1. How much of the surface is covered with fat bloom?

Question 2. What is the bloom density of the (worst) bloomed area?

Table 8.2 shows the different answers that the experts could give to the two questions above. Then, a mean score was computed, with a weight of 70% and 30% for the first and second question, respectively.

From the fat bloom scores that were computed from the experts' answers, an order relation was established. The pralines were ranked from the lowest to the highest score. The ranking of a set of 13 pralines submitted to the experts' assessment is shown in Table 8.3. Note that a low (resp. high) score corresponds to pralines with a low (resp. high) degree of fat bloom.

TABLE 8.2 Conversion of the Experts' Answers into Scores

Answer	Score
0%	0
Between 1% and 24%	1
Between 25% and 49%	2
Between 50% and 74%	3
Between 75% and 99%	4
100%	5

TABLE 8.3 Experts' Ranking

Ranking	Praline Identification	Score
1	874	0.0
1	428	0.0
3	049	2.6
4	282	3.2
5	856	4.6
5	602	4.6
7	560	5.2
8	456	6.6
9	749	6.8
10	313	7.4
11	034	8.0
12	746	8.6
13	283	10.0

The objective of this study is to apply the models presented in this work to represent the preference relation previously established. Note that the modeling was done on photographs of pralines and not on the real pralines, whereas the panel evaluated the real pralines. Pictures of 13 pralines with different degrees of fat bloom were available. Each of these pictures represents a praline standing on a uniformly gray-colored background. The pralines were placed on a sloping support (30°) in a white reflection box, and the pictures were taken from above with a fixed camera and light. Figure 8.9 shows an example of such a praline picture.

FIGURE 8.9 Praline picture.

FIGURE 8.10 Example of preprocessed praline image.

Before applying the models, some image preprocessing is performed. The background of the pictures as well as the sides of the pralines have to be removed. Therefore, the top part of the praline picture is extracted. Figure 8.10 shows an example of a preprocessed picture. This is a disk having a diameter of 730 pixels.

Note that the original color pictures have been converted to a gray scale. Therefore, fat bloom evaluation of a picture will be done by using the value of each pixel of the praline on the gray scale. Such a scale represents the intensity of each pixel, from the lowest value 0, which corresponds to a black pixel, to the highest, 255, which represents a white pixel. The intermediate values between 0 and 255 represent shades of gray.

8.3.3 Representation of Preferences over a Set of Praline Pictures

In the context of maps comparison, decision maps are partitioned into polygons or pixels that are assessed on a scale composed of a finite number of ordered categories. In the context of pralines comparison, pictures are partitioned into pixels that are assessed on the gray scale. Each shade of gray constitutes a category. It is assumed that a decision maker has a preference relation defined on the set of pictures. Here, it is assumed that the preference relation is given by the order relation established from the experts' scores. More precisely, for any two praline pictures A and B, we have $A \succ B$ (resp. $A \sim B$) if and only if the score given by the experts to the praline represented in picture A is lower than (resp. equal to) the score of the praline represented in picture B. The objective is to try to restore the experts' ranking in Table 8.3.

8.3.3.1 Model 1: Expected Utility

First, the basic expected utility model is applied to see what is obtained as a ranking. In the context of pralines comparison, the model is expressed as

$$u(A) = \frac{1}{N} \sum_{i=0}^{255} n_i(A) u_i, \tag{8.5}$$

where A denotes a praline picture, N is the total number of pixels in the region of interest (i.e., in the disk of diameter length equal to 730 pixels), $n_i(A)$ is the number of pixels (in the region of interest of picture A) having i as gray scale value, and u_i is the utility associated with gray scale value i. To compute the value of $u(A)$ for any picture A, the value of u_i has to be set. Because there is no *a priori* information regarding the differentiation of gray scale levels, u_i is defined here as an affine transformation of the gray scale value i, setting $u_i = 255 - i$. In this way, the higher the utility value $u(A)$, the better the state of praline A. Note that a value of 0 on the gray scale corresponds to color "black" and 255 to color "white." Consequently, the utility associated with a black pixel is 255 and the utility associated with a white pixel is 0. Table 8.4 shows the ranking and the scores obtained using the expected utility model (8.5) and Figure 8.11 shows the praline pictures ranked from the best to the worst according to this model. The last column recalls the experts' ranking.

The rankings given by the model and by the experts, match exactly on the best and the four worst pralines. There are some small rank differences. For instance, picture 456 is ranked in the ninth position by the model and in the eighth position by the experts. Larger differences are also observed as, for example, picture 049, which is ranked in the seventh position by the model and in the third position by the experts, or picture 749, which is ranked in the fifth position by the model and in the ninth position by the experts.

TABLE 8.4 Utility Model Scores

Ranking	Image	Score	Experts' Ranking
1	874	197.86	1
2	282	191.37	4
3	428	191.23	1
4	560	190.96	7
5	749	190.30	9
6	602	189.83	5
7	049	188.45	3
8	856	186.54	5
9	456	181.55	8
10	313	178.15	10
11	034	175.18	11
12	746	157.37	12
13	283	152.86	13

FIGURE 8.11 Chocolate images ranked from the best (top left) to the worst (bottom right) according to the expected utility model.

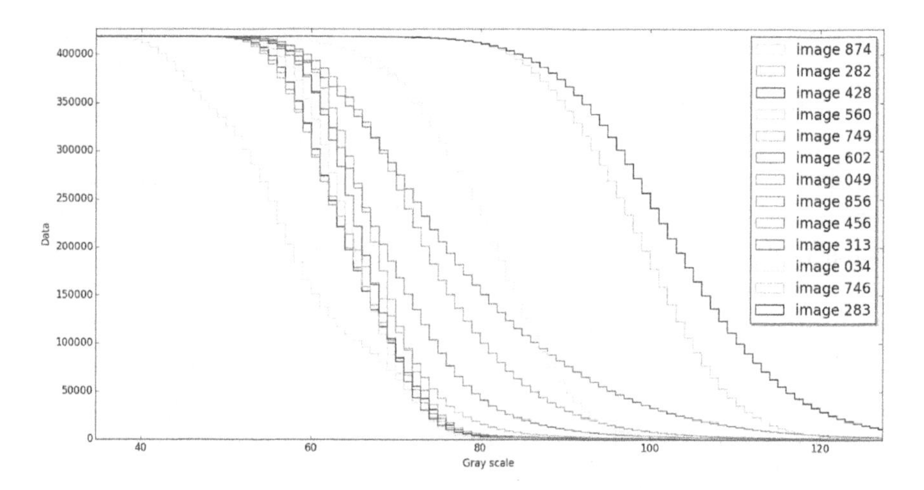

FIGURE 8.12 Cumulative histogram of the data.

Figure 8.12 shows cumulative histograms of the shades of gray for the 13 pralines. For each value on the gray scale, the corresponding point on the curve represents the number of pixels showing a whiter color. Curve 874 clearly corresponds to the praline in the best state because its histogram quickly decreases. Histograms 313, 034, 746, and 283 correspond to pralines that seem to be in a definitely worse state as compared to the others. Regarding the rest of the pralines sample, the curves intertwine and it is more delicate to determine which one is better than another.

The expected utility model, with utilities that are linear recodings of the gray scale, gives results that differ from the experts' ranking. Finding polynomial utilities could give a better ranking. To this end, an UTA-poly method (Sobrie et al., 2018) is applied. Basically, the program is fed with the ranking provided by the experts, the number of pixels assigned to each color and the degree of the polynomial. The program returns utilities u_i that restore the expected ranking, if it finds some. Here, the UTA-poly method was not able to search for a polynomial with a degree greater than 2, and even in this case, it returned a polynomial of degree 1 (our linear utilities).

8.3.3.2 Model 2: Choquet Integral

The expected utility model assumes that the only thing that matters is the frequency distribution of the pixels among the gray scale values; the spatial disposition of the pixels on the praline surface has no influence. We remind that the experts' ranking is based on their answers to two questions: the proportion of the surface covered with fat bloom and the bloom density of the worst bloomed area. The expected utility model only takes into account the number of pixels assigned to each gray scale value. This seems to be more related to the first question. The second question seems to be more related to "contiguity" in the sense that the bloom density is higher when white pixels are grouped together. That is the reason why we used the model presented in Section 8.2.2, which is able to take into account the contiguity of pixels. According to the experts, for the same amount of whitish spots, a consumer would certainly prefer to have scattered small white spots than a large white spot.

Assume that two pixels are contiguous if and only if they have a common "border." Figure 8.13 illustrates the pixels contiguous to a pixel, p. Also, assume that only contiguous pixels do interact (i.e., there are no

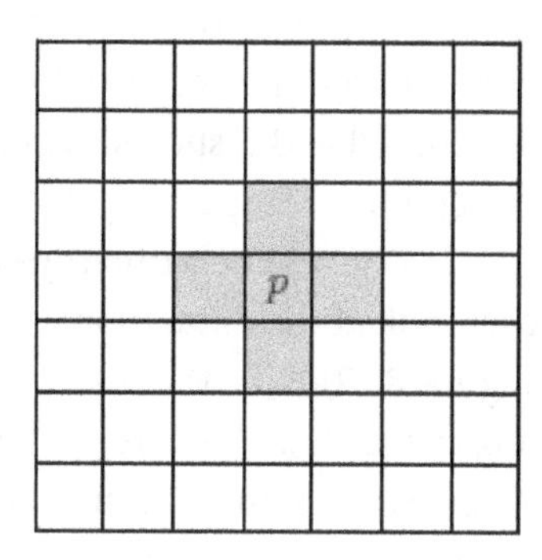

FIGURE 8.13 Contiguous pixels.

longer range interactions). In this case, the Choquet integral of the utility of a praline picture A reads as follows:

$$C_m(u,A)=\sum_{s\in S} m(\{s\})u(s)+\sum_{\substack{s,t\in S\\ \{s,t\}\text{ contiguous}}} m(\{s,t\})\min(u(s),u(t)). \quad (8.6)$$

Giving the same importance to each pixel and to each pair of contiguous pixels, we can set $m(\{s\})=\alpha$ and $m(\{s,t\})=\beta$. Consequently, expression 8.6 becomes

$$\begin{aligned} C_m(u,A) &= \alpha\sum_{s\in S} u(s)+\beta\sum_{\substack{s,t\in S\\ \{s,t\}\text{ contiguous}}} \min(u(s),u(t)) \\ &= \alpha\sum_{i=0}^{255} n_i(A)u_i+\beta\sum_{i=0}^{255} m_i(A)u_i, \end{aligned} \quad (8.7)$$

where $n_i(A)$ denotes the number of pixels having i as gray scale value in the region of interest of picture A, $m_i(A)$ denotes the number of contiguous pairs of pixels such that the worst one (i.e., the whitest one) has i as gray scale value. The parameters α and β are two constants satisfying $\alpha N+\beta M=1$ with N the total number of pixels in the region of interest, and M the total number of contiguous pairs of pixels. Note that u_i is still defined as an affine transformation of the gray scale value i, (i.e., $u_i = 255 - i$). The first term in (8.7) is similar to the expected utility model. The second term gives an advantage (if β is positive) or a disadvantage (if β is negative) to contiguous white pixels as compared to scattered ones. Indeed, consider two images having the same distribution on the gray scale, as in Figures 8.7 and 8.8. In case of a positive β value, the second term is larger for images in which transition between white and black pixels seldom occurs (i.e., where white pixels are grouped together and so are black pixels). On the opposite, with negative β values, a scattered distribution of white pixels over the praline surface is favored. Table 8.5 shows the contribution of each of the two terms in the Choquet integral separately. It can be observed that the interaction alone (second term) gives the same ranking as the main effect (first term). Having a closer look at the data, more precisely, comparing the histogram of the data (Figure 8.12) with the histogram of the interactions (Figure 8.14), it can be observed that they are very similar. This explains why the interaction alone gives the same ranking as the main effect. Also, this fact implies that it is impossible to obtain an interesting combination by using positive β coefficient. This is no problem because contiguous

TABLE 8.5 Contribution of the First and Second Terms in the Choquet Integral

Image	$\frac{1}{N}\sum_{i=0}^{255} n_i(\cdot)u_i$	$\frac{1}{M}\sum_{i=0}^{255} m_i(\cdot)u_i$
874	197.86	196.61
282	191.37	190.12
428	191.23	189.95
560	190.96	189.68
749	190.30	188.88
602	189.83	188.54
049	188.45	187.15
856	186.54	185.10
456	181.55	179.78
313	178.15	175.76
034	175.18	173.76
746	157.37	155.80
283	152.86	150.73

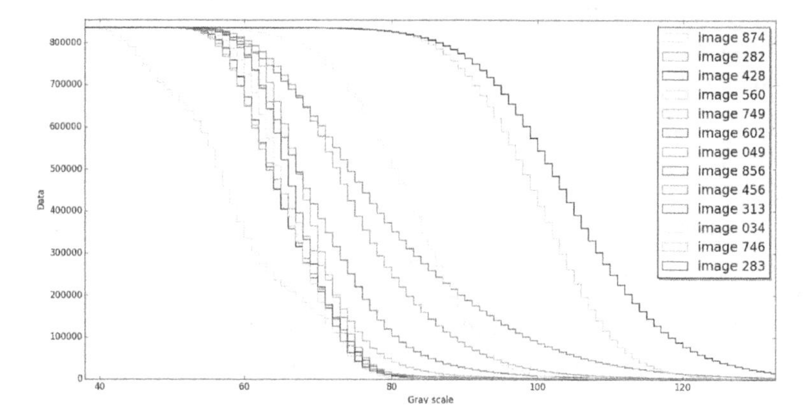

FIGURE 8.14 Cumulative histogram of the interaction term.

white pixels should be avoided (i.e., negative values of β should be used). Unfortunately, setting β to a negative value does not affect the ranking obtained by only considering the main effect, either. Table 8.6 shows the scores obtained with β equal to its lower bound (negative).[1]

Comparing the results in Table 8.4, obtained with the expected utility model, and the results in Table 8.6, obtained using a Choquet integral

[1] For values of β smaller than this bound, expression (8.7) is not guaranteed to be nondecreasing in the u_i's.

TABLE 8.6 Choquet Integral Scores with Nearest Neighbor Interaction Penalizing Contiguous White Pixels

Ranking	Image	Score
1	874	196.62
2	282	190.12
3	428	189.95
4	560	189.68
5	749	188.87
6	602	188.54
7	049	187.15
8	856	185.09
9	456	179.82
10	313	175.77
11	034	173.81
12	746	155.83
13	283	150.75

penalizing contiguous white pixels, it can be observed that the rankings are the same. This means that taking into account pairs of pixels having a common border does not alter the ordering obtained using the simple expected utility model. That is why we tried to take into account more complex interactions, among which:

1. Taking into account pairwise interactions with larger neighborhoods as illustrated in Figure 8.15.
2. Taking into account interactions with more than two pixels. In particular, we studied interactions inside square groups of pixels as illustrated in Figure 8.16.

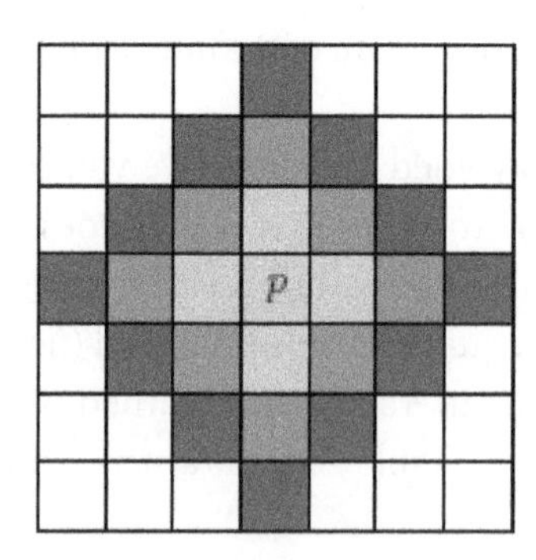

FIGURE 8.15 Pairwise interactions with larger neighborhood.

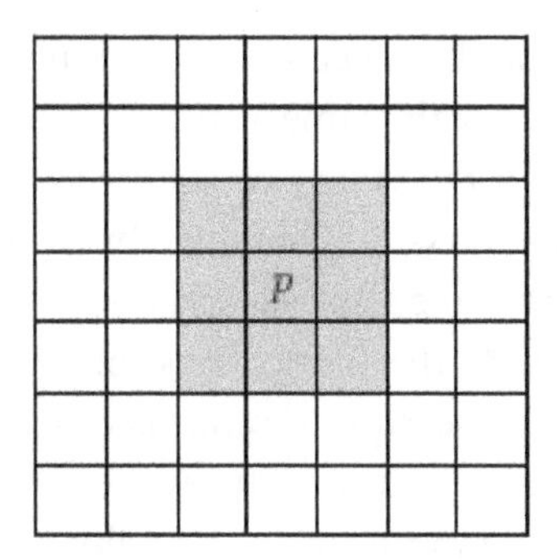

FIGURE 8.16 Square interactions.

3. Using a multilinear model (see Krantz et al., 1971, Chapter 6) to represent the interactions. Instead of considering the worst evaluation among a set of pixels, we consider the product of the pixels' utility.

Those models still yield the same ranking as the one obtained with the expected utility model.

8.4 DISCUSSION

It was observed, in Section 8.3.3, that all the models provide the same ranking which is different from the experts' one. Such differences may be because of the inadequacy of models that do not capture the features on which the experts rely to make their assessments. Besides the models, there are two main circumstances that could explain these differences.

1. Before using the models, some image preprocessing was applied, potentially resulting in loss of relevant information. In particular, the aspect of the border of the upper part of the praline as well as its side may have an influence on the experts' judgments. These parts were removed from the pictures.

2. A single picture was used for each praline. The pictures were obtained by placing the pralines on a sloping support and they were taken from above. The experts did not work with these pictures but directly by looking at the real pralines. This allowed them to place the pralines in whatever position they want. It turns out that, in practice, this has a big influence on fat bloom detection. Depending on the position and also on the light, fat bloom can sometimes be observed or not.

Regarding the latter point, the experts have been subsequently asked to evaluate fat bloom on the basis of the same pictures as these at our disposal. First, they were asked to answer the same two questions as they did with the real pralines. On this basis, a ranking is obtained by using the rule described in Section 8.3.2. The experts were also asked to directly rank the pictures from the best (least bloomed) to the worse (most bloomed). Table 8.7 shows a summary of our ranking, the experts' ranking based on the pralines scores, the experts' ranking based on the pictures scores and the experts' direct ranking of the pictures. From this table, the following can be observed.

1. All the rankings put the same praline in the first place. Also they identify the same four worst (but not exactly in the same order). The conclusion would be that the first one is definitely better than the others and that the four last ones are worse than all the others. This was already observed in Figure 8.12.
2. The experts' ranking obtained from the scores given to the pictures is definitely different from the one obtained by directly ranking the pictures. A possible explanation is that, according to the experts, sample 874 clearly was the best sample and that the last 4–5 samples were the easiest to rank, but the samples in between were difficult to rank as there was almost no fat bloom visible. This difficulty is likely to have generated variation in the experts' judgments that we notice in Table 8.7.

TABLE 8.7 Comparison between Pralines Ranking and Pictures Ranking

Images	Our Ranking	Experts' Ranking Based on Pralines Scores	Experts' Ranking Based on Pictures Scores	Experts' Direct Ranking of the Pictures
874	1	1	1	1
282	2	4	5	3
428	3	1	2	2
560	4	7	6	5
749	5	9	8	7
602	6	5	4	4
049	7	3	3	6
856	8	5	7	8
456	9	8	9	9
313	10	10	11	11
034	11	11	10	10
746	12	12	13	12
283	13	13	12	13

3. The ranking obtained with the models is more similar to the experts' direct ranking of the pictures than to the others. Comparing both pictures direct rankings, it can be observed that there are never more than two places between the position of a picture, whereas there are up to four places when comparing the ranking, obtained with the models, with the experts' ones based on the scores. To be more formal, the Kendall tau distance between our ranking and the experts' can be computed. This metric counts the number of pairs (i, j) such that i is placed before j in one ranking and j is placed before i in the other ranking. Comparing the ranking, obtained with the models, with the experts' ranking based on pralines scores (resp. with the experts' ranking based on pictures scores and the experts' ranking of the pictures), the Kendall tau distance is equal to 11 (resp. 10 and 5). Because there is a total of $\frac{13(13-1)}{2}$ pairs, there are $\frac{11}{78} \approx 14\%$ (resp. $\frac{10}{78} \approx 13\%$ and $\frac{5}{78} \approx 6\%$) pairs that differ between the corresponding rankings.

From this extensive study, the following conclusions may be drawn.

- Pralines can be rank ordered with a certain reliability with respect to fat bloom on the basis of standardized preprocessed photographs and the frequency distribution of gray pixels in the praline pictures.
- Using a model penalizing the concentration of whitish pixels in contiguous zones does not contribute to a better matching with the experts' ranking.
- The experts' ranking is sensitive, to a certain extent, to the assessment protocol (answering two questions on a specific scale) and to working with praline pictures instead of real chocolates. A further research path consists of trying to capture the features assessed by the experts while answering two questions by using appropriate models. The first question, for instance, explicitly refers to the frequency distribution of the pixels in shades of gray as does our first model (expected utility).

REFERENCES

Aissi, H., S. Chakhar, and V. Mousseau. GIS-based multicriteria evaluation approach for corridor siting. *Environment and Planning B: Planning and Design*, 390(2): 287–307, 2012.

Brison, V., *Spatial decision aiding models for maps comparison*. PhD thesis, Université de Mons, Faculté polytechnique, Mons, Belgique, 2017.

Chakhar, S., *Cartographie Décisionnelle Multicritère: Formalisation et Implémentation Informatique*. PhD thesis, Université Paris Dauphine, France, 2006.

Chateauneuf, A., and J.Y. Jaffray. Some characterizations of lower probabilities and other monotone capacities through the use of Möbius inversion. *Mathematical Social Sciences*, 17: 263–283, 1989.

Joerin, F., *Décider sur le territoire. Proposition d'une approche par utilisation du SIG et de méthodes d'analyse multicritère*. PhD thesis, Ecole Polytechnique Fédérale de Lausanne, Suisse, 1998.

Krantz, D.H., R.D. Luce, P. Suppes, and A. Tversky. *Foundations of Measurement, Volume 1: Additive and Polynomial Representations*. Academic Press, New York, 1971.

Malczewski, J., GIS-based multicriteria decision analysis: A survey of the literature. *International Journal of Geographical Information Science*, 20(7): 703–726, 2006.

Malczewski, J., Multiple criteria decision analysis and geographic information systems. In *Trends in Multiple Criteria Decision Analysis*, pp. 369–395. Springer, New York, 2010.

Metchebon Takougang, S.A., *Contribution à l'aide à la décison en matière de gestion spatialisée. Etude de cas en management environnemental et développement de nouveaux outils*. PhD thesis, Université de Mons, Faculté polytechnique, Mons, Belgique, 2010.

Metchebon, S.A.T., M. Pirlot, S. Yonkeu, and B. Some. *Evaluation and decision models with multiple criteria: Case studies*, chapter assessing the esponse to land degradation risk: The case of the Loulouka catchment basin in Burkina Faso, In. *International Handbooks on Information Systems*, pages 341–399. Springer, New York, 2015.

Metchebon, S.A.T., V. Brison, and M. Pirlot. Two models for comparing decisional maps. *International Journal of Multicriteria Decision Making*, 3, 129–156, 2013.

Nopens, I., I. Foubert, V. De Graef, D. Van Laere, K. Dewettinck, and P. Vanrolleghem. Automated image analysis tool for migration fat bloom evaluation of chocolate coated food products. *LWT—Food Science and Technology*, 41(10): 1884–1891, 2008.

Reinke, S.K., S.V. Roth, G. Santoro, J. Vieira, S. Heinrich, and S. Palzer. Tracking structural changes in lipid-based multicomponent food materials due to oil migration by microfocus small-angle x-ray scattering. *ACS Applied Materials & Interfaces*, 7(18): 9929–9936, 2015.

Roy, B., and D. Bouyssou. *Aide Multicritère à la Décision: Méthodes et Cas*. Economica, Paris, France, 1993.

Sobrie, O., M. Pirlot, and F. Joerin. Intégration de la méthode d'aide à la décision ELECTRE TRI dans un système d'information géographique open source. *Revue Internationale de Géomatique*, 23(1): 13–38, 2013.

Sobrie, O., N. Gillis, V. Mousseau, and M. Pirlot. UTA-poly and UTA-splines: Additive value functions with polynomial marginals. *European Journal of Operational Research*, 264(2): 405–418, 2018.

CHAPTER 9

Multi-Criteria Decision Planning with Anticipatory Networks to Ensuring the Sustainability of a Digital Knowledge Platform

Andrzej M.J. Skulimowski

CONTENTS

9.1 INTRODUCTION

Anticipatory networks (ANs) are one of new approaches to multi-criteria decision making that is strongly combined with researching the future [1–4]. This chapter presents some recent extensions of the AN-based decision-modeling tools [2] in the context of their applicability in multi-criteria decision making (MCDM) and sustainable planning with multiple criteria. ANs provide constructive algorithms to computing nondominated solutions that comply with the anticipatory preference structure. Specifically, we will show how to apply the anticipatory decision-making principles to select best-compromise solutions of present-time multi-criteria sustainable strategy planning problems based on feasible visions of future optimization tasks and their outcomes. This construction merges judgmental forecasts derived from a multi-round Delphi survey [5,6], the statistical forecasts of future decision problem parameters, and a simulation of future decision selection procedures. The latter may be used to derive decision rules and preference structures to be applied as a base of holistic anticipation with an evolving AN.

The main object of study in the AN theory is the system of decision units linked by causal and anticipatory feedback relations. To each active decision unit there is associated a decision maker and a problem to be solved. A sequence of future decisions linked by causal relations is termed an admissible (decision) chain [2]. The assumption that future decision makers always strive to admit Pareto-optimal decisions allows the analyst to filter out admissible chains with at least one dominated solution. Those that remain will be termed *rational*. Given a reference point q_i for a decision problem O_i [7], a solution sequence ensuring that given causal successors $D_{j1},\ldots,D_{j(i)}$ of the decision maker D_i associated to O_i can select decisions equal to or better than their anticipated reference points $q_{j1},\ldots,q_{j(i)}$ will be termed *sustainable*. In general, a sequence of causally linked decisions that fulfill predefined conditions concerning the impact on solutions to be chosen by decision makers in the future will be termed *anticipatory decision chains*. A background of the AN-based sustainability theory is provided in Section 9.3.

Analysis of future circumstances, parameters and their mutual relations with, for example, morphological analysis [8] and AN-based assessment processes allows the decision maker to select a subset of scenarios out of the computed future visions. These scenarios will be regarded as normative [9] so that an AN-based backcasting [10] can be run while the parameters of normative scenarios correspond to ideal or aspiration objective values describing the overall strategy implementation. To find an appropriate decision chain that ensures reaching the reference levels of criteria one can apply the Algs. 1, 1a, and 2 provided in [2]. They are backcasting-oriented, that is, seeking a solution of the current problem starting from certain acceptable solutions admitted at the planning horizon. Such decision sequences fulfill the above definition of anticipatory chains. However, anticipatory chains often do not exist. If a normative scenario is infeasible (i.e., no decision chains are able to reach their goals), one can find a decision sequence included in an admissible chain that yields closest, in a predefined sense, criteria values to the future vision characteristics of this scenario. The best-compromise scenario thus found describes the most desired future and starts from a present-day best-compromise decision [11]. Another approach consists in finding all feasible decision chains starting from the current problem and remaining anticipatory for a maximum number of future decision makers or for a longest period possible. Then, best acceptable decision sequences in terms of a proximity measure can be specified. Such sequences correspond to the elementary exploratory scenarios [9,12] of the overall network evolution. In this chapter, the latter approach, based on exploratory scenarios rather than normative, has been admitted, which corresponds to the situation where no strict planning horizon can be specified.

More information about ANs and their MCDM applications are provided in the next section. The reader interested in this method of modeling decision consequences may find further information in [2,3], as well as in [4], and to a broader context in [5]. A bibliographic survey of AN-related ideas can be found in [4].

The previously presented anticipatory decision model constitutes a methodological background to build an exploitation strategy for an innovative knowledge repository developed within an EU Horizon 2020 project [13]. Strategy building for an information system is a relatively new type of strategic planning activity. Actually, few papers have been published on this subject, and most of them are devoted to the systems of digital libraries, such as open repositories for e-learning or building cohesive digital learning repository systems [14,15]. Generally, in the context of strategic management, knowledge repositories can be regarded as virtual enterprises with strong dynamic capabilities [16] and with a salient specificity. This is why a dedicated methodology had to be elaborated based on the ANs theory and an earlier experience with digital repositories serving foresight support systems. This application of the ANs theory is presented in Section 9.4, while a real-life example referring to strategic planning of the knowledge platform is analyzed in Section 9.5. The findings presented in this paper are summed up in the concluding Section 9.6.

9.2 ANTICIPATORY NETWORKS AS A STRATEGIC PLANNING TOOL

One of the origins of ANs is the theory of anticipatory systems. By definition, an *anticipatory system* in the Rosen's [17] sense makes its decisions based on a future model of itself and of its outer environment. The anticipatory behavior of decision makers represented as AN nodes fulfills the Rosen's definition of anticipatory system as well as its further extensions by other researchers [18].

As it has been already mentioned in the introduction, AN-based models can formalize both multi-stage multicriteria forward planning and multicriteria backcasting. ANs generalize earlier anticipatory models of decision impact in multi-criteria problem solving [1] and constitute an alternative decision model to utility or value function estimations and to diverse heuristics. An AN is a directed multigraph with nodes modeling the decision problems and edges modeling causal influence, anticipatory feedback, and information exchange relations between the decision makers. A node in an anticipatory network with no predecessors is termed the *starting node*. It models the present-time multi-criteria decision problem, termed the *starting problem*. The other nodes model decisions in future multi-criteria problems to be solved by the same or different decision makers. If the starting node is not unique then this AN models group decision making. However, such network topologies will not be considered in this chapter. The edges of the first kind model the dependence of decision problems on the solutions made to some earlier problems. There may be several causal relations and the corresponding edge classes in one network. A more formal definition of an AN is provided later in this section (cf. Def. 9.8).

The fundamental assumption of AN theory says that the impact of previous decisions manifests in imposing some constraints on subsequent decisions. Following the introductory remarks, this assumption is represented as a graph of a causal relation linking the following multi-criteria optimization problems

$$(F_i : U_i \rightarrow IR^{N_i}) \rightarrow min(\theta_i),\ i=0,1,\ldots,k, \tag{9.1}$$

where F_i and U_i are vector objective functions and sets of admissible decisions, respectively. It is assumed that for each i, the convex cone θ_i defines the partial order $\leq_{\theta i}$, which may equivalently describe constraints on trade-off coefficients at each i-th node. The 0-th node is the starting problem. As usual in the MCDM theory, agents responsible for decision making will be referred to as decision makers. The optimization problems of type (9.1) together with implicitly assigned agents responsible for selecting and implementing decisions will be the nodes of ANs modeling actual decision-makers' behavior.

Definition 9.1

The decision units, where the decision maker D_i solves the problem (9.1) during the period of time $[t_{i-}, t_{i+}]$ will be termed (multi-criteria) *optimizers* and denoted by O_i.

The sets of nondominated (Pareto-optimal) points and nondominated values at each node are defined respectively as

$$P(U_i, F_i, \theta_i) := \{u \in U_i : [\forall v \in U_i : F_i(v) \leq_{\theta i} F_i(u) \Rightarrow F_i(v) = F_i(u)]\},$$

$$F_i P(U_i, \theta_i) := F_i(P(U_i, F_i, \theta_i)).$$

Nondominated points are *solutions* to each of the problems (9.1) and *candidate solutions* to the multi-criteria strategy selection problem in an AN. To formulate the latter, we need to define two relevant relations between optimizers.

Definition 9.2

The optimizers O_i and O_j for some i, j, $0 \leq i, j \leq k$, $i \neq j$, are in a causal domain dependence relation r if O_i precedes O_j in the temporal order, by definition, $t_{i+} \leq t_{j-}$, and there exists a non-trivial and non-constant multifunction φ_{ij} defined on U_i with values in U_j, $\varphi_{ij}: U_i \rightarrow 2^{U_j}$, such that the choice of a decision u_i in the problem O_i sets the decision scope of the problem O_j to $\varphi_{ij}(u_i)$.

Recall that a multifunction φ_{ij} is non-trivial if its values are different from U_j and from the empty set.

The additional constraints on the decision choice in a subsequent optimization problem depend more explicitly on the outcomes of the decision made at O_j, i.e., on $F_i(u)$ rather than on u. This is why in previous papers [1–3] we analyzed the situation where φ_{ij} is a superposition of F_i and certain multifunction defined in the criteria space. However, to keep the notation simpler, in this chapter we will not separately analyze the dependence on criteria values. Moreover, we will consider only one domain dependence relation, so without an ambiguity r can be termed just a *causal relation*. We will also assume that

1. If $O_{i1}\ r\ O_j$ and $O_{i2}\ r\ O_j$ then the decision scope of the problem O_j is the set $\varphi_{i1,j}(u_{i1}) \cup \varphi_{i2,j}(u_{i2})$, where u_{i1} and u_{i2} are any decisions admitted by O_{i1} and O_{i2}, respectively.
2. If O_i and O_j are in a causal relation r defined by φ_{ij}, then there is no *temporal conflict* between the decision makers at O_i and O_j; by definition this means that

$$\forall\ i,j: O_i\ r\ O_j \forall u,v \in U_i [F_i(v) \leq_{\theta_i} F_i(u) \Rightarrow \varphi_{ij}(v) \setminus \varphi_{ij}(u) \neq \varnothing]. \tag{9.2}$$

The first assumption means that the relation r is *enabling*, that is, it allows D_j to choose a solution from any set defined by a causal predecessor of O_j. In general, the combination of causal predecessor requirements concerning the decision scope of O_j may be any Boolean expression binding the sets $\varphi_{i1,j}(u_{i1}),\ldots,\varphi_{in,j}(u_{in})$, cf. [3]. The condition (2) in the second assumption means that a better decision made by D_i at O_i cannot restrict the decision scope of D_j at O_j without compensating this with new selection options. Remark that a stronger requirement, the *temporal cooperation* condition, namely

$$F_i(v) \leq_{\theta_i} F_i(u) \Rightarrow \varphi_{ij}(u) \subset \varphi_{ij}(v),$$

may appear too strong for modeling the decision process in the application case studied in Sections 9.4 and 9.5.

The decision choice in problems embedded in an AN bases on a constructive analysis of causal relations that link the outcomes of an MCDM problem in (9.1) with their future consequences expressed as constraints imposed on the scope of future decisions. These constraints are defined as values of the linking multifunctions [1]. Besides the causal relations between optimizers, another major element of the anticipatory decision theory is the deployment of anticipatory information in the network solution process. Specifically, we assume that when solving the current problem (1) with $i = 0$, the decision maker takes into account anticipated results of solving some future decision problems. The decision maker D_i selects an own decision in such a manner that the restrictions imposed by the multifunctions φ_{ij} defined by D_i on the decision scope of its successor D_j facilitate selecting at O_j a decision desired by D_i. Of course, such causal relation between O_j and the problem being just solved must exist. Distinguishing more and less preferred solutions to one or more problems solved by the causal successors of O_i defines thus an additional preference structure among the elements of U_i. Namely, a solution $u_{i1} \in U_i$ that guarantees the satisfaction of more D_i's desiderata concerning the choice of future solutions than u_{i2} is preferred to u_{i2}. The satisfaction of D_i's desiderata may also be partial, in such a case it makes sense to select the decision that guarantees that the future solutions satisfy the desiderata to a higher extent. This new preference structure supplements the original preference relations expressed by the cones θ_I in (9.1). It will be termed the *anticipatory preference relation* in U_i (APR). Before providing a strict definition of APR (cf. Def. 9.5), we will define first the anticipatory feedback between optimizers.

Definition 9.3

The additional preferences of D_i concerning the selection of solution to a future problem O_j by D_j will be termed the *anticipatory feedback relation* (AF) between O_j and O_i. An AF is expressed by the second type of directed edges inserted into the graph of a causal relation r. They start at the nodes where the decisions are to be influenced and end at certain nodes that correspond to the influencing multi-criteria decision problems. An AF between O_j and O_i will be denoted by $f_{j,i}$ and so will be labeled the AF edges. ■

Observe that an AF makes sense only if O_i can influence the choice of a decision at O_j. Otherwise the additional preferences of D_i cannot be expressed by the choice of a decision at O_i. In this chapter we will consider two complementary types of AFs defined in Defs. 9.4 and 9.7 below.

Definition 9.4

Suppose that G is the graph of a causal order relation r between optimizers $O_0,\ldots,O_k$ and that the optimizer O_i precedes O_j in r. If the decision maker D_i at O_i defines the preference relation $a'(f_{j,i})$ among the elements of the decision set U_i then O_j and O_i are in the *primary anticipatory feedback relation* (*primary AF*) denoted pAF. The relation $a'(f_{j,i})$ is termed an *anticipatory order.* ■

Here, we will assume that D_i defines the set $V_{j,i} \subset F_j(U_j)$ termed the *set of desired outcomes* at O_j, based on the parameter forecasts of O_j and own wishes as regards the decision to be made by O_j. Then, the *strict anticipatory order* $a_s(f_{j,i})$ in U_j is defined for every u_p, $u_q \in U_j$ in the following way:

$$u_p\ a_s(f_{j,i})u_q \Leftrightarrow^{df} [F_j(u_p) \in V_{j,i} \cap F_j P((U_j,\theta_j) \wedge F_j(u_q) \notin V_{j,i} \cap F_j P((U_j,\theta_j))]$$
$$\vee [F_j(u_p) \in V_{j,i} \wedge F_j(u_q) \notin V_{j,i}] \tag{9.3}$$

The decisions u_p and u_q are *anticipatorily indiscernible*, which will be denoted by $u_p \neg a_s(f_{j,i})\ u_q$, if, by definition, neither $(u_p, u_q) \in a_s(f_{j,i})$ nor $(u_q, u_p) \in a_s(f_{j,i})$. It is easy to see that $\neg a_s(f_{j,i})$ is an equivalence relation with at most three non-empty abstraction classes

$$\alpha_1 := F_j^{-1}(V_{j,i} \cap F_j P((U_j,\theta_j)), \alpha_2 := F_j^{-1}(V_{j,i}), \cap U_j, \text{ and}$$
$$\alpha_3 := U_j \setminus F_j^{-1}(V_{j,i} \cup F_j P(U_j,\theta_j)).$$

Moreover,

$$a'(f_{j,i}) := \neg a_s(f_{j,i}) \cup a_s(f_{j,i}) \tag{9.4}$$

is an anticipatory order in U_j. Thus, the sets $V_{j,i}$ define the pAF between O_j and O_i. We assume that once the set $V_{j,i}$ is defined, D_j knows it and respects the preferences of D_i by striving to select a decision from the set

$$W_j := F_j^{-1}(V_{j,i} \cap F_j P(\varphi_{ij}(u_i),\theta_j)) \cap W_{a,j},$$

where u_i is the influencing decision selected by D_i and $W_{a,j}$ is the subset of U_j which elements fulfill to the maximal extent (which will be defined later on) the AFs with the end node O_j. Obviously, W_j may be empty; then the choice of a solution at O_j should be performed according to a certain rule defining the priority relation between the native preference structure of D_j expressed by F_j and θ_j, the own anticipatory preference structure expressed by $W_{a,j}$, and the satisfaction of D_i's wishes expressed by $V_{j,i}$. Here, we admit the preemptive priority rule identical to the order the preference requirements have been listed in the previous phrase. Additionally, a regularizing distance function similar to the function h in Alg. 1a in [2] may be applied.

If there are $m>1$ AFs starting at O_j and the intersection $V_{j,i1} \cap \ldots \cap V_{j,im}$ is empty then D_j's decisions may follow a certain a priority order satisfied by the individual AFs. This order will be constructed making use of the above defined $a'(f_{j,i})$. It may also take into account the AFs ending at O_j.

Observe first that the relation $a'(f_{j,i})$ generates in a natural way an order among the solutions to the problem O_i, which is formally defined below.

Definition 9.5

Suppose that the optimizers O_j and O_i are nodes in a graph G of the causal relation r with an influencing multifunction φ_{ij} and that they are linked additionally by a primary AF. Then the *anticipatory preference relation* (APR) between the elements u_{i1}, u_{i2} of the decision set U_i is the following order $a(f_{j,i})$ induced on U_i by $a'(f_{j,i})$ from U_j:

$$u_{i1}\ a\left(f_{j,i}\right)\ u_{i2} \Leftrightarrow^{df} \min\{m : \varphi_{ij}\left(u_{i1}\right) \cap \alpha_m \neq \phi\} \leq \min\{n : \varphi_{ij}\left(u_{i2}\right) \cap \alpha_n \neq \phi\}, \tag{9.5}$$

where $\alpha_m, 1 \leq m \leq 3$, are the above defined abstraction classes of the anticipatory indiscernibility $\neg a_s(f_{j,i})$. ■

It is easy to see that from the definitions of $\alpha_m, 1 \leq m \leq 3$, and $a(f_{j,i})$ that it follows that (9.5) holds if and only if

$$\forall 1 \leq m \leq 3 \forall v_1 \in \varphi_{ij}\left(u_{i1}\right) \cap \alpha_m\ \forall v_2 \in \varphi_{ij}\left(u_{i2}\right) \cap \alpha_m :\ v_1 a'\left(f_{j,i}\right)\ v_2.$$

If the inequality in (9.5) is sharp then the corresponding relation will be termed *strong anticipatory preference relation* (sAPR) and the condition above should be replaced by

$$\forall 1 \leq m \leq 3 \forall v_1 \in \varphi_{ij}\left(u_{i1}\right) \cap \alpha_m\ \forall v_2 \in \varphi_{ij}\left(u_{i2}\right) \cap \alpha_m :\ v_1\ a_s\left(f_{j,i}\right)\ v_2.$$

We will say that an APR is *plausible* if $\alpha_2 \neq \phi$. Given an O_i, there may be more than one anticipatory feedbacks $f_{j,i}$ between O_j and O_i, for different $j \in J(i) \subset \{1,\ldots,k\}$ such that O_j is causally dependent on O_i. The requirement that all of them are simultaneously satisfied is equivalent to the fact that all APRs related to the anticipatory feedbacks $f_{j(p),i}$ are plausible. Admitting the notation $J(i):=\{j(1),\ldots,j(p)\}$, this requirement may be formulated as the following condition

$$\exists u \in U_i : (\varphi_{i,j(1)}(u) \cap \alpha_{m(1)} \neq \phi) \wedge \ldots \wedge (\varphi_{i,j(p)}(u) \cap \alpha_{m(p)} \neq \phi),$$
$$\text{with } m(j) \leq 2 \text{ for } j \in J(i), \tag{9.6}$$

where $\varphi_{I,j(q)}$, $q=1,\ldots,p$, are multifunctions that define the scope of admissible decisions at O_j. They are provided either in the definition of direct causal dependence between O_i and O_j or are calculated as the union of superposed multifunctions along all causal chains $c \in C(i,j)$ linking O_i and O_j,

$$\varphi_{ij} := \cup_{c \in C(i,j)} \varphi_{c(p_c),j} \circ \ldots \circ \varphi_{i,c(1)},$$

where p_c denotes the number of elements (length) of the chain c.

As we already mentioned, it may happen that one or more from these intersections is empty i.e., the *anticipatory preference relation* generated by the Boolean product of all AFs at O_i is not plausible. In this case one can define various relaxation rules to specify an element of U_i, which satisfies the anticipatory feedbacks at O_j to a maximum extent possible. The rule admitted in this chapter selects an $u \in U_i$ with the highest count of satisfied conditions $\varphi_{ij}(u) \cap \alpha_{m(j)} \neq \phi$. It is defined as follows:

Definition 9.6

If the optimizer O_i is in an AF with O_j for $j \in J(i)$ then the order

$$u_{i1}\ a(f_{j,i})u_{i2} \Leftrightarrow^{df} \#\{j \in J(i) : \varphi_{ij}(u_{i1}) \cap \alpha_{m(j)} \neq \phi\} \leq \#\{j \in J(i) :$$
$$\varphi_{ij}(u_{i2}) \cap \alpha_{m(j)} \neq \phi\}, \tag{9.7}$$

for any $m(j) \leq 2$, $j \in J(i)$, defines the compromise solution selection rule at O_i allowing the D_i to choose an element that satisfies the highest number of AFs. ■

The set of $u \in U_i$ satisfying the condition (9.6) or the above surrogate rule (9.7) with the maximum count will be denoted by Z_i. It will be used to define the secondary AF in Def. 9.7.

Definition 9.7

Optimizers O_j and O_i such that O_i r O_j are in the *secondary anticipatory feedback* relation (*secondary AF, sAF*) if the decision maker D_i at O_i wishes D_j could select a decision satisfying all anticipatory feedbacks (primary or secondary) with the end node O_j. If there are no secondary AFs ending at O_j, then one of the conditions (9.6) or (9.7) is satisfied with the set Z_j playing the same role as $V_{j,i}$ in pAF. If at least one sAF ends at O_j, Z_j must be defined with recursive application of this definition along all paths of sAFs ending at O_j until a pAF is encountered in all of them. ■

Thus, the set of solutions at O_j solicited by the decision maker D_i at O_i coincides with the intersection of all sets of the form $F_j^{-1}(V_{k,j})$, where $V_{k,j}$ defines an anticipatory feedback (primary or secondary) at O_j. If Def. 9.7 is supplemented by a stipulation that D_j always selects a solution that satisfies the surrogate decision rule based on (9.7), the secondary AF remains well defined even in case the intersection of $F_j^{-1}(V_{k,j})$ for all *kÎJ(j)*, is empty. Let us note that this rule may be defined in many possible ways, different from the proposed (e.g., there may be a relevance hierarchy of AFs at O_i so that D_i selects first the decision from the abstraction classes corresponding to the higher-ranked AFs).

Let us note that the secondary AF between O_j and O_i in the networks of optimizers may be equivalent to the primary AF between the same optimizers with the set Z_j playing the same role as $V_{j,i}$ in the primary feedback. Therefore, from commonsense reasons, we will assume that primary AF between O_j and O_i excludes the secondary one between the same optimizers, albeit this would be formally allowed. Because the definition of the sAF is recursive, it may require finding the decision subsets fulfilling the sAF for all decision units O_k such that O_j precedes O_k in the causal order. In [2] and [3], secondary AFs were not distinguished because, by an assumption made there, all successors' anticipatory requirements were taken into account at each decision node. This is why the computation of anticipatory decisions started there from the nodes with no predecessors in an AF.

A formal definition of an AN of optimizing decision units with both kinds of anticipatory feedbacks can now be formulated in the following way.

Definition 9.8

A connected directed multidigraph with the starting node O_0 that consists of a digraph of at least one causal relation r between optimizers and includes at least one additional primary or secondary anticipatory feedback relation linking O_0 with another node in the network will be termed an *anticipatory network* (of optimizers). ▪

Furthermore, one can take into account an information exchange relation H between decision units that need not be connected by a causal relation. By definition, $O_i\, H\, O_j$ iff the decision maker at the decision node O_i knows the decision made currently or to be made, by O_j. This type of AF has been termed *induced anticipatory feedback* [3]. An example of an AN with an induced anticipatory feedback is presented in Figure 9.1.

To sum up, the analysis of an AN consists in appropriate simultaneous handling of both types of relations, causal and anticipatory. Usually, this analysis is supplemented by forecasts or exploratory scenarios [9] regarding future decision problem parameters. There may be

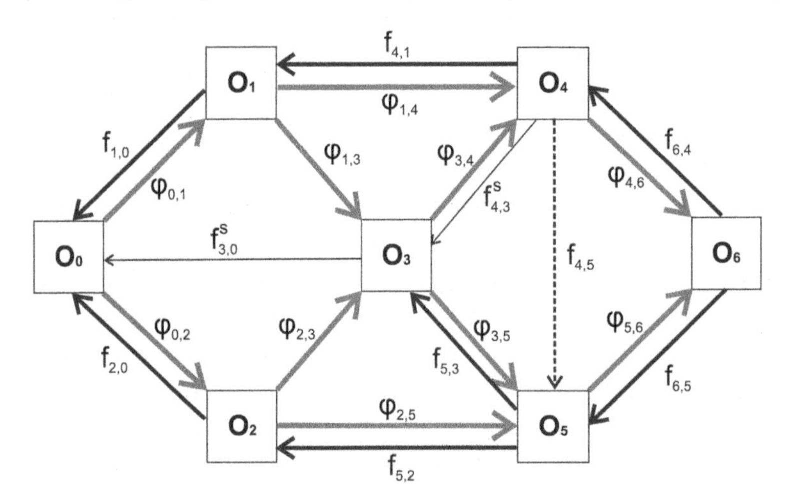

FIGURE 9.1 An example of an anticipatory network of decision units: the nodes O_i, $i = 0,\ldots,6$, denote multi-criteria decision units, *broader* arcs denote causal influence relations and temporal order between the decision units, *thinner dark* arcs denote anticipatory feedbacks indicating which future decisions are relevant to a decision maker preceding the source node of an AF in the temporal order (bold) or indicating the relevance of secondary AF satisfaction at a future node (*thin edges*, here $f^s_{3,0}$ and $f^s_{4,3}$). The dotted edge $f_{4,5}$ denotes the induced AF.

defined two additional preference structures. The first one over the set of future decision problems to determine the order the future decision-makers' preferences are taken into consideration. The other preference structure is defined over the set of anticipatory feedbacks and determines the priority they are taken into account when modeling the decision problem solution at the target or starting node of these feedbacks. Due to the recursive character of secondary AF, for any two secondary feedbacks at causally dependent nodes O_j and O_i such that $O_j \, r \, O_i$, the second from the preference structures given here must assign a higher priority to the secondary AF at O_i. In [2] it has been shown how the additional preference structure consisting of both supplementary relations can enable the decision makers to confine the sets of admissible nondominated decisions at the starting node of anticipation denoted there by O_0.

All these relations represented in an AN form a complex information model, where the decision maker associated with the problem acquires various forecasts, embeds them into the models of the problems to be solved by future decision makers and their interrelations, and makes a decision him- or herself.

9.3 THE MULTI-CRITERIA SUSTAINABILITY STRATEGY SELECTION PROBLEM

The aim of building a strategy for the digital enterprise may be manifold, from ensuring a resilience in different future external circumstances, to reaching a specific long-term market target. In addition, the use of the term "strategy" is often colloquial even in scholarly publications. One of the reasons for this is the existence of many non-equivalent definitions that circulate in business, political, and research contexts [15], in game theory, and decision planning. However, there are some common points of most, if not all of them, namely:

1. The definition of the strategy objects and their relevant aspects,
2. The goal setting in the temporal context, from short- to long-term, and their prioritization,
3. Long-term action planning to reach the goals, involving an acquisition and use of forecasts or scenarios,

4. Modeling the uncertainty resulting from the future states of nature, from an approximate character of the strategy object evolution rules, or from actions of other actors,

5. Designing the decision algorithms to respond to the uncertainties and reach the goals in an optimal way.

The components of the generic strategy building have been also admitted in this chapter to elaborate on the digital knowledge repository exploitation strategy. It has been assumed that the goal attainment will be evaluated quantitatively by the set of strategic criteria F. The strategic goals have to be attained at a prescribed moment in the future, called the *planning horizon*, with further requirements to be observed at intermediate moments. The overall strategy implementation requires a concerted activity of multiple decision makers. We will assume that their decisions are stretched in time, and together with relations between the problems and decision makers, they are embedded in an AN.

A similar problem, yet to a lesser extent, touches upon the definition of *sustainability*. Following the well-known Brundtland's definition [19], this can be meant here as *ensuring the decision-making conditions for future generations that are not worse than those of ourselves.* We will also refer to the notion of digital sustainability, cf. [20,21] which covers additionally technological and information society aspects. For the sake of a mathematically sound sustainability modeling, we will define here the *generations* of decision units and *sustainability chains.*

Definition 9.9

A subnetwork S of an AN that consists only of optimizers and such that

1. The causal component r of S is a chain with the starting optimizer O_i and the final one O_j, $i < j$,
2. The optimization criteria functions $F_k := \{F_{k1},\ldots,F_{kp(k)}$ in the optimizers O_k, $i \le k \le j$, contain a non-empty subset $H := \{H_1,\ldots,H_p\}$, which is identical for all optimizers in S,
3. For each O_k in S different from O_i there exists an AF $f_{k,i}$ between O_k and O_i defined with the set $V_{k,i} := \{u \in U_k : F_k(u) \le_{\theta k} v_k\}$, where v_k is the *reference level* for F_k and all v_k are non-increasing along S,

4. The primary and secondary AFs form a multi-chain (any connected union of chains with the same nodes) in the reverse direction than r, that is, from O_j to O_i; if a primary AF is defined between O_k and O_{k+1} with $V_{k+1,k}$ and W_{k+1} defines sAF at $O_{k+1,}$ then $W_{k+1} \subset V_{k+1,k}$.

will be termed a *sustainability chain*. The decision makers D_k associated to O_k for $i \leq k \leq j$, will be termed the *generations* of D_i. The second part of condition 4 means that the scope of sustainable decisions planned by D_i for D_k, $k=i+1,\ldots,j$, is broader than any later AFs defined by D_k. ■

An AN will be called *sustainability-oriented* if it contains at least one non-trivial (i.e., with at least two optimizers) sustainability chain.

Making use of these notions, we will formulate the sustainable strategy selection problem embedded in an AN:

Problem 9.1 (Sustainable Strategy Selection)

Let G be an AN where constraints U_i, criteria F_i and preference structures may depend on external environmental variables e_i that can admit only a finite set of values (forecasts or scenario parameters). Furthermore, let $v_{i1},\ldots,v_{in}$ be the reference vectors for the criteria $F_{i1},\ldots,F_{in}$ for a certain sub-network $G_1 := \{O_{i1},\ldots,O_{in}\}$ of G. Let $H := (H_1,\ldots,H_p)$ be the set of additional strategic planning criteria and $h_j := (h_{j,1},\ldots,h_{j,p})$ the corresponding reference levels active at j-th optimizers in G_1 for $1 \leq j \leq n$. The vectors $v_{i1},\ldots,v_{in}$ and $h_1,\ldots h_n$ define the sets

$$V_{j,i} := \left\{ v \in IR^{nj+p} \left(F_j, H \right) \left(v \right) \leq_{\theta' j} \left(v_{ij}, h_j \right) \right\},$$

where n_j is the dimension of the vector criteria function F_{ij} at the optimizer O_{ij}, $\theta_j' := \theta_j \times IR_+^p$, and i point out the immediate predecessors of O_j in S. Then the primary AF relations $f_{j,i}$ with $V_{ij,i}$ are defined according to Def. 9.4. Assume moreover that each chain in G_1 with the modified criteria $F_{ij}' := \left(F_{ij}, H \right)$ is a sustainability chain (Def. 9.9). Then a collection of multi-criteria problems $O_{ij}' := (U_{ij}, (F_{ij}, H), \theta_{ij}')$, *for* $j = 1,\ldots,n$, linked with the causal relation r inherited from G and AFs defined with $V_{j,i}$, form a *sustainability anticipatory network*. The *sustainability strategy* is an algorithm solving any chain in the anticipatory subnetwork G_1, that is, providing compromise anticipatory solutions in terms of Def. 9.6 to all optimizers along any chain in G_1 for a given combination of the external variables e_i. The compromise values of the strategic criteria $H_1,\ldots,H_p$ at each optimizer in G_1 yielded by this algorithm will be called the *strategy assessment*. ■

Observe that the above problem differs from those formulated in [2], where an anticipatory compromise solution was sought at the initial decision node only. To sum up, ANs that can be applied to modeling and solving real-life sustainable strategy selection problems can be built with the following information about the future [2]:

- Exploratory scenarios or forecasts concerning the parameters of future decision problems represented by the decision sets U, criteria F, the preference models P of each future decision maker. Their attitudes toward anticipatory planning (rational/partly rational/irrational) are also taken into account.
- The causal dependence relations r linking the nodes in the network.
- The anticipatory feedback relations pointing out which future outcomes are relevant when making decisions at end nodes of AFs and specifying the anticipatory feedback conditions (AFC).

The solution process of Problem 9.1 aims at selecting a compromise at each optimizer in G_I and getting a satisfactory strategy assessment for a possibly large—in terms of a probabilistic measure—subset of plausible external parameters e_j. Taking them into account may be accomplished either in a simulation procedure, aimed at yielding a robust decision rules for each O_i, or—in a less formal way—within a roadmapping process [22], where the results of a repeated solution scheme are discussed with the strategic stakeholders. In both cases a single-instance of the solution process may use the existing AN solving algorithms, according to the following statement.

Proposition 9.1

The preference information contained in an AN G and criteria H may be applied in Algs. 1, 1a, and 2 from [2] to filter the plausible exploratory scenarios, playing thus the role of a sustainable strategy in terms of Problem 9.1 and yielding a compromise strategy assessment, provided that:

- All future agents whose decisions are modeled in the network are rational (i.e., they make their decisions so that they are consistent with their preference structures).
- An agent can assess to which extent the outcomes of causally dependent future decision problems are desired. This relation is described

by multifunctions linking present-time decisions with future constraints and preference structures.

- The first two assumptions and the corresponding assessments are transformed into decision rules to be applied at the current decision problem. It is also assumed that the latter affects the outcomes of future problems in a known way.
- There exists a relevance hierarchy H_1 in the network G; usually the more distant in time an agent is (modeled by a node in G), the less relevant the choice of her or his solution.
- There exists a family of relevance hierarchies H_2 of anticipatory feedbacks in the network G; usually the more distant in time the starting node of an AF is, the less relevant the choice of his or her solution. ■

Both relevance hierarchies allow first, the strategy stakeholders, and then the decision makers at the optimizers to derive a partial order that can serve to define the sequence of operations in decision selection algorithms. The decision-making process in an AN is equivalent to filtering the set of all causal chains in the network that are determined by sequences of admissible decisions made along a causal path. This procedure is valid for the networks of cooperative agents that look for a Pareto-optimal strategy as well as for hybrid networks [3] that include antagonistic games with Nash equilibria.

9.4 A REAL-LIFE APPLICATION: THE MULTI-CRITERIA STRATEGY SELECTION FOR A DIGITAL KNOWLEDGE PLATFORM

As an application example, where the ANs turned out to be a particularly well-suited modeling tool, we will present the multi-criteria planning problem of an innovative digital knowledge platform (cf. [13]). Multiple criteria at the initial problem and so-called future-generation problems model are related to the financial sustainability, social impact and benefits, risks and threats, and the strategic position of the platform among similar knowledge-based web tools. The "future-generation" decision makers are included in the network as a new class of nodes and model the sustainability of the platform. Specifically, the causal relations modeling the influence of an "earlier-generation" decision maker on the future management, exploitation, and development decisions made by a "next generation" must fulfill a set of conditions ensuring a sufficient scope and other preconditions of future decisions. In this example, the initial problem corresponds to the decisions of the coordinating body of the consortium conducting the research project aiming the platform development, and the "next generations" correspond to the future management of the platform. The other decision makers are responsible for the user community building and technological development decisions. The anticipatory structure of this problem is shown in Figure 9.2.

A research consortium consisting of the Coordinator (C) and several research units R_i, $i = 1,\ldots,I$, is responsible for the design and implementation of the digital knowledge platform (MP). It is also responsible for the deployment of the platform by "Use case operators" (UC), although the latter will become independent after the implementation project is finished.

As it is depicted in Figure 9.2, the platform may serve multiple user communities, although for a simplicity of the numerical example presented here, only two communities will be taken into account. The first one will group young researchers from one or more higher-learning institutions, the other will consist of financial auditors from a large auditing company. These communities are presented in Figure 9.2 and consist of "Generic institutional users", $GU_1,\ldots,GU_m$ surrounded by a cloud of final users (FU_i). Community of each type will be supported by "Use case operators" $\left(UC\left(j\right),\, j=1,\ldots,J\right)$, each of them serving possibly multiple "Institutional

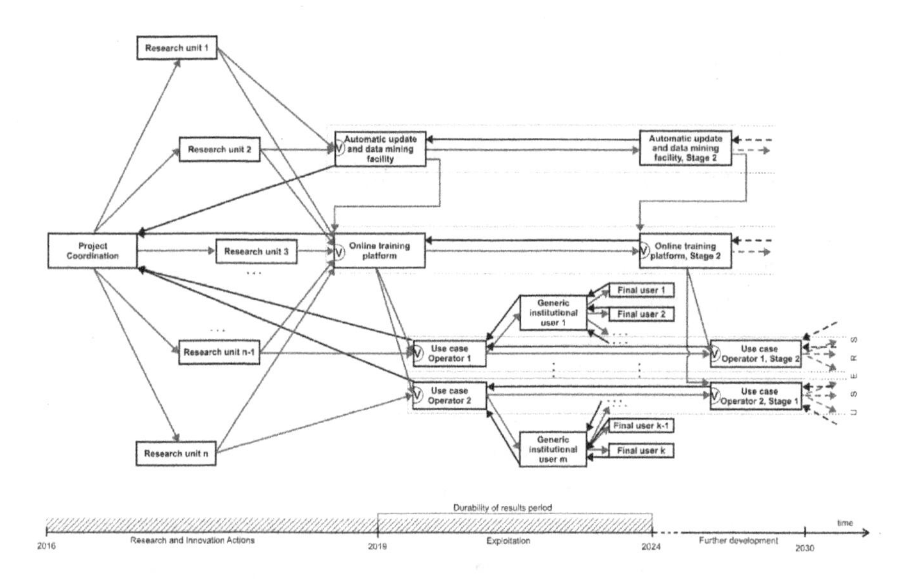

FIGURE 9.2 An anticipatory network modeling the strategic decisions concerning a digital knowledge platform.

users" (GU). The end users (individual or institutional) may benefit from training and information services provided either by the Institutional users or by the "Use case operators" directly.

The data and technological (digital) sustainability of the platform will be ensured by technological upgrades of the platform itself and of the "Automatic data mining facility" (DMF). The latter will be responsible for automatic data gathering from the outer environment and other repositories and for feeding them to the platform. The anticipatory decision-making problem considered in this paper is motivated by the strategy planning problem for an academic knowledge platform which operates under the assumption that all decision makers involved cooperate directly. This is why an induced feedback related to a contemporary decision will be handled by an information exchange between the involved decision makers rather than by forcing one of them to make a specific decision with an induced feedback. Consequently, induced feedbacks do not occur in the AN presented in Figure 9.2.

This general functioning scheme modeled by an AN constituted a base to build a sustainability strategy for the platform. The first main goal of this strategy is to ensure the undisturbed performance of the platform according to the terms and specifications elaborated during the project over the period of at least 5 years after its end. So is the durability of the project results defined. The second goal is the long-term viability and sustainability of the platform, which may need to undergo major technological refurbishing after several years of operation.

The information and communication technology (ICT) roadmapping merged with the AN framework has been chosen as the best-suited methodology to construct both components of the strategic plan for the platform: a middle-term strategy ensuring the durability of results and a long-term sustainability strategy. As evidenced by earlier applications, roadmapping is capable of merging technological and business development aspects of research results in an efficient way [22]. In addition, publicly available information technology foresight can be taken into account when modeling future deployment environments of the platform. Similarly, demand forecasts may be used to optimize customer acquisition based on the priority ranking of marketing and development activities. The latter refer to implementing updated services and functionalities. An initial technological roadmapping has been performed during the platform-development stage to elicit the relations between different agents involved in platform functioning, including the user communities and to forecast their parameters. The latter include specifically the constraints and criteria of future decision problems to be solved by each agent. Roadmapping process has been concerted with constructing the AN model when generating nondominated platform exploitation strategies. Best-compromise strategy selection and visualization have been performed with the roadmapping diagrams as well as with ANs.

The overall sustainable strategy building is a complex collaborative process involving all project partners and external experts. It assumes a continuous knowledge acquisition from heterogeneous information sources, its managing, processing, and sharing among the end users based on the experience gathered. All use cases have been given equal attention during the strategy building, involving elements of business planning such as the New Product Development and Market Placement roadmapping approach [22]. However, most uses of the platform are assumed to be not-for-profit, open access for the worldwide research community.

9.5 THE STRATEGIC PLANNING PROCESS

The anticipatory network model (ANM) allowed the strategy building team to structure the temporal relations among the individual components of the internal platform design, construction, and management, the community of its users, and external social, research, technological, and economic environment. Then, the scope of best-compromise decisions is determined for each internal unit within the model so that the decisions made ensure the sustainability of the platform in all five aspects presented. According to the general rules of building the ANMs, the model consists of the elements shown in Figure 9.2.

The decision-making units are specified in Table 9.1 together with their admissible actions. To comply with the real-life case, we assume that there are two use case operators, UC(1) and UC(2). The sustainability is modeled as the relations between two generations of the platform and the data mining facility, DMF (both identified with their development stages) and of two user communities. It is assumed that any further generation of these units depend on the predecessor's decisions in the same way as the second generation on the first one. Furthermore, for each use case only one generic institutional user and one generic final user are considered in the strategy computing. This simplification does not restrict the number of users provided that the user communities are homogeneous at each level and the anticipatory expectations from each user in the same use case are identical. In such a situation the anticipatory feedbacks between institutional users and use case operators are defined in the same way and the same action of UC(i) satisfies all feedbacks from institutional users connected to this use

TABLE 9.1 Decision-Making Units and Their Admissible Actions

Unit No.	Unit Description	Admissible Actions	Notation
1	[R_i] Research units $R_1,\ldots,R_I$ taking part in the implementation of the platform and the coordinator C	$e_{i,Y}$—actual efforts spent by R_i on the implementation of the unit Y, where Y stands for MP (the platform), DMF (data mining facility), or UC(1), UC(2) (use case operators); $e_{i,Y} = e_{i,Y}(0) + \Sigma_{1 \le k \le i}\, \delta_{k,i}(Y)$, where $e_{i,Y}(0)$ denote resources initially assigned to perform the activities concerning the implementation of Y before any changes are made. The values of $e_{i,Y}(0)$ and all $\delta_{k,i}(Y)$ are predetermined within a group decision procedure by the consortium led by C, where each of the R_i is involved, but it performs the activities agreed as soon as a consensus decision is made. Thus, R_i are dependent decision units and the influence of R_i on Y is a functionally dependent on U_C	$U_{R(i)} = \{e_{i,MP}, e_{DMF}, e_{i,UC(1)}, e_{i,UC(2)}\}$; $e_{i,Y}>0 \Leftrightarrow R_i$ causally influences Y
2	[DMF] Data mining facility (*N* generations)	The data mining facility is assigned an initial endowment $d(0)$. It allocates $d_n(1)$ to its own current operation costs and $d_n(2)$ to the DMF development. Additionally, it transfers $d_n(3)$ to its next generation successor. The number of such allocations is equal to the number N of generations considered, i.e., $n = 1,\ldots,N$ (in this example $N = 2$). The values of $d_n(1)$ and $d_n(2)$ determine the scope and quality of data fed to the platform MP in the following way: $$D_{n+1} := \{D_{1,n+1},\ldots,D_{\nu(n+1),n+1}\} := \xi\left(d_n(1),d_n(2)\right),\ n=1,\ldots,N,$$ where $\xi: = (\xi_1,\ldots,\xi_{\nu(n)})$ is certain vector function transforming the allocation of expenses $d_n(1)$, $d_n(2)$ into the next-step operational features *of* DMF denoted by D_{n+1}	$U_{DMF,1}: = \{(d_1(1),d_1(2),d_1(3))_\sigma, s\},\ldots,$ $U_{DMF,N} = \{(d_N(1),d_N(2),d_N(3))_\sigma, s\}$, where σ is an index ordering all possible allocations such that $d_1(1) + d_1(2) + d_1(3) = d(0)$ for the initial generation of DMF, while for each subsequent one $d_n(1) + d_n(2) + d_n(3) =$ $d_{n-1}(3) + p_{n-1}(4)$, for $n = 1,\ldots,N$. The funding $p_{n-1,4}$ comes from MP; it is defined in p. 3

(Continued)

TABLE 9.1 (*Continued*) Decision-Making Units and Their Admissible Actions

Unit No.	Unit Description	Admissible Actions	Notation
3	[MP] Online training platform (N generations)	The platform allocates its initial endowment $p(0) := p_1(0)$ to its own current operation $p_n(1)$ and development $p_n(2)$, as well as transfers $p_n(3)$ to its next generation and $p_n(4)$ to the next generation of the data mining facility. N such allocations are considered. The amount of funds available for each n-th generation of MP may increase on $q_n := \Sigma_{j=1,\dots J}\, q_n(j)$ – the fees transferred from the $(n-1)$th generation of use case operators UC(j), $j = 1,2$; $n = 2,\dots,N$ (cf. unit 4 below). In addition, MP influences the UC(j) by defining the rates $r_{n,k}$ (in %) of UC(j) incomes to be transferred to the next-generation MP. Only K values of $r_{n,k}$ are possible, i.e., $k = 1,\dots,K$. The values of $p_n(1)$, $p_n(2)$ and q_n together with the data D_n fed to the platform from the DMF determine the scope, quantity, and quality of platform services available to use case operators, $S_n = \eta\left(p_n(1), p_n(2), q_n, D_n\right) = \left\{S_{1,n},\dots,S_{m(n),n}\right\},\ n = 1,\dots,N.$ The transferred amounts of $p_n(3)$ and $p_n(4)$ influence the scope of actions of next-generation MP and DMF, respectively.	$U_{MP,1} = \{(p_1(1), p_1(2), p_1(3), p_1(4))_\sigma, s\} \times \{r_{1,1},\dots, r_{1,K}\},\dots,$ $U_{MP,N} = \{(p_N(1), p_N(2), p_N(3), p_N(4))_\sigma, s\} \times \{r_{N,1},\dots, r_{N,K}\},$ with $p_n(1) + p_n(2) + p_n(3) = p(0) + q_n(1) + q_n(2)$ for $n = 1,\dots,N$. For simplicity's sake for all $n, m = 1,\dots,N$ $r_{n,i} = r_{m,I}$ and $K = 2$.

(*Continued*)

TABLE 9.1 (*Continued*) Decision-Making Units and Their Admissible Actions

Unit No.	Unit Description	Admissible Actions	Notation
4	[UC(1), UC(2)] Use case operators (*M* generations)	Each of the n-th generation use case operators allocates available funds $c_{n0}(j)$ to its own operational costs $c_n(j,1)$, to the user community building $c_n(j,2)$, and transfers $c_n(j,3)$ to its next-generation successor. N subsequent allocations corresponding to the subsequent generations of UC(j) are considered in the model. The initial endowment $c_0(j)$, $j = 1,2$, available at the first stage of the use case operation results from the project fund allocation. The next generations of UC(j) may also benefit from the fees $c_{n,f}(j)$ collected from the institutional and final users that will be split into the obligatory transfer of $q_n(j)$ to the next-generation MP. The share of $c_{n,f}(j)$ to be transferred is determined by MP. The value of $c_{n,f}(j)$ depends monotonically on $M(n,j)$.	$U_{1,UC(j)}:= \{c_1(j,1),c_{10}(j,2),c_{1UC}(j,2),c_1(j,3))$ $,\ldots, c_Q(j,1),c_{1Q}(j,2),c_{QU}(j,2);c_Q(j,3)),s\}$, with $c_k(j,1) + c_k(j,2) + c_k(j,3)) = c_0(j) + c_f(j),\ldots,$ $U_{N,UC(j)}:= \{c_N(j,1),c_{N0}(j,2),c_{n,UC}(j,2),c_N(j,3)),$ $\ldots, c_Q(j,1),c_{1Q}(j,2),c_{QU}(j,2),c_Q(j,3)),s\}$,
5	[GU] Generic institutional [IU] and Final users [FU]	Generic users GU (IU and FU) make market decisions whether to use the platform services or not based on the services offered S_n and their prices. Each use case operator is characterized by the number of its affiliated users $M(n,j)$, $j = 1,2$, at each generation $n = 1,\ldots,N$. These quantities should steadily grow according to the sustainability principle. The IU(j), or the institutions responsible for supplying the platform services to the final users, such as government agencies, nongovernmental organizations, pay for the services supplied by the use case operators. The total amount paid to each of the use case operators is equal to $c_{n,f}(j)$ that depends directly on the number of services sold, and on the number of final users $M(n,j)$. Thus, according to Eq. (9.8), it depends indirectly on the set of services provided by the UC(j) and the amount spent for promotion of the platform and user community building.	The interactions between IU and FU in terms of causal influence and anticipatory feedbacks exist as well. However, they are not included in the present model as the Management Units C and MP have no, or only a very limited influence on these relations. Thus, the decisions made by IU are only assumed optimal in terms of generating the income from FU and their sponsors and increasing the loyalty and size of the community of users.

case operator. A similar observation touches on the anticipatory feedback relation between the final users and institutional users. Namely, they point out the same set of desired actions of each final user.

A heterogeneous community of users will require different actions targeted at each of the different subpopulations that define different feedbacks and may react differently to an action of UC(i), $i = 1,2$. This leads to the formulation of a combinatorial multi-criteria optimization problem to find a sequence of UC(i) actions that satisfy all strategic user community-building criteria in a compromise way. However, the general model of sustainability is not affected by the diversity of user classes as in each use case the trade-off between the resources passed to the next generation of UC and those spent on the promotion and other supplementary services provided to the users is more relevant.

The admissible actions for each of the decision-making units are presented in col. 2 of Table 9.1.

In rows 4 and 5 in the table, we assumed that the amount of $c_n(j,2)$ determines the growth of the number of final users $M(n, j)$ according to the equation

$$M(n+1,j) = M(n,j) + \gamma\left(c_n(j,2), S_n\right), \tag{9.8}$$

where γ is certain monotonic function of $c_n(j,2)$.

The activities presented in Table 9.1 fulfill the following assumptions:

1. The service offer depends only on the factors here, that is, only a finite number of different allocations is possible, all rounded to 5% of the total amount available to the n-th generation, and that an allocation of 5% to MP allows to increase the number of services on 1 or to increase the quality of one existing service to the next competitiveness level.

2. A re-allocation of funds between different tasks within the same unit R_i is a joint decision of R_i and C. Although it is usually initiated by R_i that knows the anticipatory feedback between the project output units, the Coordinator and the Steering Committee (C) must approve it. Only those self-re-allocations will be explicitly analyzed that influence the capacities or operation outcomes of other units.

3. A re-allocation is possible only if from a forecast update it follows that the project goals will be better fulfilled if more (or less) effort is devoted to:

a. The implementation of further services or functionalities of the platform MP,

b. The implementation of the data mining facility DMF,

c. The actions of the use case operators UC(j), $j = 1,\ldots,J$.

The status-quo solution (no re-allocation) corresponds to the initial project budget and objectives.

4. Only those re-allocations that actually affect the capacities of MP, DMF, and UC(i) will be analyzed. All those remaining that handle operational risk or improve the performance of individual R_i's are not considered because the AFs between R_i's and C are omitted as well. The latter touch on the project's operational performance while the duties of R_i and the responsibilities of C are part of the project contract.
5. The number and value of services sold by the platform via commercial use case operators are functions of the services offer available and the quantity of the community of users. The parameters of Eq. (9.8) will be determined ex ante—for the first stage of planning with market research and ex-post—for the next user generations by the webometric facilities measuring the performance of the MP (quantitative characteristics of MP use) and money flow (annual revenues).
6. The compliance of the user community building by the use case operators and technological expectations of the users will be assured by taking into account the AFs of the users at both user levels: individual and institutional. For brevity's sake, we skip here detailed characteristics of AN causal and anticipatory edges.

Finding the optimal balance of further development and user community-building actions for the network optimizers' tasks according to the decision problem structure presented in Figure 9.2 has been further performed according to the Algs. 1, 1a, and 2 in [2] and with the anticipatory preference relation $a(f_{j,i})$ defined by (9.7). The comparability of both approaches was assured by admitting an appropriate form of the regularizing functions h_i in Alg. 1a in [2], which are defined as $h_i(x) := \#J(i) - \#\{j: x \in V_{j,i}\}$ for each i such that O_i is an ending node of an AF. The first method was chosen for the solution variant with a predefined strategic planning horizon, while the second method was used in all other cases. Both methods yielded similar results when the relations $a(f_{j,i})$ were applied in a network trimmed to the planning horizon applied in Alg. 1a. It turned out that the major trade-off problem to be solved is related to the allocation of resources between the units responsible for the further development of MP and the use case operators UC responsible for the community building. This led us to formulate a subordinated social impact modeling and optimization problem with the criteria describing the structure of the user community and the satisfaction of users. This social impact model details the decision problems presented in rows 4 and 5 of Table 9.1 and will be solved separately. The resulting hints have been passed to the project [13] partners as a part of the platform's Final Exploitation Strategy.

9.6 CONCLUSIONS

Building an exploitation strategy for a knowledge repository with multiple goals, criteria, and contexts turned out to be a new type of strategic-planning activity. The earlier work on development strategies for classical book libraries turned out to a great extent outdated, while the newer papers devoted to the exploitation planning of digital libraries, e-learning course repositories, or other learning management systems did not cover the overall range of services offered by the learning platform developed within the project [13]. For related work the reader is referred to, for example, the reports on open repositories for e-learning [23] or on building cohesive digital learning repository systems in Thailand [14] or Ecuador [24]. The manifold aspects of the platform functioning, specifically its social, economic, and technological features resulted in a multi-criteria problem formulation of the strategy-building process. The solution process that uses the AN formalism can additionally benefit from multilevel multi-criteria optimization methods [18], in particular in a hybrid networks with competition and leader-follower gaming behavior.

The future relations and values of parameters governing the platform exploitation taken into account during the strategy building result from forecasts performed within a dedicated Delphi survey [5]. Their updates may be elicited during periodic roadmapping organized prior to an end of an interim planning horizon. The recent state-of-the art in the platform development as well as the most up-to-date knowledge on the future project research environment will be taken into account during each road-mapping exercise. The resulting strategic plan contains virtually all activities to be undertaken to assure the sustainability of the platform. It complies with the coordinator and research units individual goals and exploitation plans.

This strategy will make it possible to attain the ultimate goal of the platform's exploitation, that is, to make it available as a training tool for young researchers, doctoral students, and public administrators. The user community will grow using the existing cooperation networks centered on current international projects and student research organizations. The platform can also be used for research purposes and can enhance existing forecasting or Foresight Support Systems and knowledge bases. All agents involved in the operation of the platform will cooperate on user group development, implementing their individual exploitation plans, and bringing further cooperation partners, whereas the input from use case providers will play a key role in defining new target user groups. Outranking procedures to prioritize the community-building activities can also be applied. A special attention should be given to the platform presence in social media, focusing on research-oriented and professionally oriented social networks to pertain a permanent inflow of new platform users. The best practices in building Internet communities and an appropriate platform positioning principles are taken into account. Recommending tools and facilities have been built into the platform, so as to enable the community growth based on the snowball principle. All they will contribute to reaching the sustainability goals.

It should be emphasized that the models presented in this chapter benefit from a synergy with other analytical decision support methods and the corresponding IT tools such as a group model building, and online multiround Delphi management systems (cf. [5], www.moving-survey.ipbf.eu). However, the implementation of the AN-based decision model can also be used as a stand-alone decision support system. Finally, it should also be mentioned that AN-based decision-support systems similar to those presented in this chapter have been also recently applied to strategic or cooperation planning in other organizations, for example to:

- Select technological investment strategies for a software company. One of the technologies is a class of decision-support systems embedded into an innovative ERP system developed by the company. For each software evolution scenario, the AN allows the management to better assess the system's development costs and the corresponding decision impact on other technological areas within the company.

- Building a strategy to align R&D investment projects at the regional or country level to a sustainable development strategy based on smart specializations. A real-life example refers to the strategy planning for a regional Knowledge Transfer Center providing ICT support to innovative SMEs, where recommendations to the R&D policy makers will also be derived with an anticipatory model.
- Constructing multi-criteria decision-making models for swarms of autonomous vehicles [4].

REFERENCES

1. A.M.J. Skulimowski (1985). Solving Vector Optimization Problems via Multilevel Analysis of Foreseen Consequences. *Found Control Engineering* vol. 10(1), 25–38, 1985, https://www.researchgate.net/publication/228804191_Solving_Vector_Optimization_Problems_via_Multilevel_Analysis_of_Foreseen_Consequences.
2. A.M.J. Skulimowski (2014). Anticipatory Network Models of Multicriteria Decision-Making Processes. *International Journal of Systems Science* 45(1) 39–59, doi:10.1080/00207721.2012.670308.
3. A.M.J. Skulimowski (2016). The art of anticipatory decision making. In: George A. Papadopoulos et al. (ed.). KICSS 2014: 9th International Conference on Knowledge, Information and Creativity Support Systems, Limassol, Cyprus, November 6–8, 2014, Proceedings, Advances in Intelligent Systems and Computing, vol. 416, pp. 17–35, Springer International Publishing AG. doi:10.1007/978-3-319-27478-2_2.
4. A.M.J. Skulimowski (2019). Anticipatory Networks. In: Poli R. (eds.) *Handbook of Anticipation*. Springer International Publishing AG, Cham, Switzerland, pp. 995–1030, https://doi.org/10.1007/978-3-319-91554-8_22.
5. A.M.J. Skulimowski (2017). Expert Delphi Survey as a Cloud-Based Decision Support Service, *IEEE 10th International Conference on Service-Oriented Computing and Applications SOCA 2017*, 22–25 November 2017, Kanazawa, Japan. IEEE, Piscataway, NJ, pp. 190–197, http://ieeexplore.ieee.org/document/8241542.
6. P. Tapio (2003). Disaggregative policy Delphi: Using cluster analysis as a tool for systematic scenario formation. *Technological Forecasting and Social Change* 70(1), 83–101. https://doi.org/10.1016/S0040-1625(01)00177-9.
7. A.M.J. Skulimowski (2019). *Selected Methods, Applications, and Challenges of Multicriteria Optimization*. Series: Monographs, Vol. 19, Committee for Automation and Robotics of the Polish Academy of Sciences, AGH Publishers, Kraków.
8. T. Arciszewski (2018). Morphological Analysis in Inventive Engineering. *Technological Forecasting and Social Change* 126, 92–101. https://doi.org/10.1016/j.techfore.2017.10.013.

9. L. Börjeson, M. Höjer, K.H. Dreborg, T. Ekvall, G. Finnveden (2006). Scenario types and techniques: Towards a user's guide. *Futures* 38, 723–739, https://doi.org/10.1016/j.futures.2005.12.002.
10. J. Quist, P. Vergragt (2006). Past and future of backcasting: The shift to stakeholder participation and a proposal for a methodological framework. *Futures* 38, 1027–1045. https://doi.org/10.1016/j.futures.2006.02.010.
11. M. Godet (2001). *Creating Futures—Scenario Planning as a Strategic Management Tool.* Economica, London, UK.
12. T. Comes, N. Wijngaards, B. Van de Walle (2015). Exploring the future: Runtime scenario selection for complex and time-bound decisions. *Technological Forecasting Social Change* 97, 29–46, http://dx.doi.org/10.1016/j.techfore.2014.03.009.
13. MOVING project web site: www.moving-project.eu.
14. S. Chunwijitra, P. Tummarattananont, S. Laokok, K. Krairaksa, C. Junlouchai, W.N. Chai, C. Wutiwiwatchai (2015). The strategy to sustainable sharing resources repository for massive open online courses in Thailand. In: *12th International Conference on Electrical Engineering, Electronics, Computer, Telecommunications and Information Technology (ECTI-CON)*, June 2015, IEEE CPS, doi:10.1109/ECTICon.2015.7206980.
15. P. Kotler, R. Berger, N. Bickhoff (2010). *The Quintessence of Strategic Management.* Springer-Verlag, Berlin, Germany, p. 133, doi:10.1007/978-3-642-14544-5_1.
16. D.J. Teece (2007). Explicating dynamic capabilities: The nature and microfoundations of (sustainable) enterprise performance. *Strategic Management Journal* 28(13), 1319–1350. https://doi.org/10.1002/smj.640.
17. R. Rosen (1985). *Anticipatory Systems–Philosophical, Mathematical and Methodological Foundations.* Pergamon Press, London, UK; 2nd ed. Springer, 2012.
18. I. Nishizaki, M. Sakawa (2009). *Cooperative and Noncooperative Multi-Level Programming.* OR/CS Interfaces Series Vol. 48, Springer, Berlin, Germany.
19. G.H. Brundtland (1989). Global Change and Our Common Future. *Environment: Science and Policy for Sustainable Development* 31(5), 16–43, https://doi.org/10.1080/00139157.1989.9928941.
20. K. Bradley (2007). Defining Digital Sustainability. *Library Trends* 56(1), 148–163. doi:10.1353/lib.2007.0044.
21. Y. Mizuno, Y. Kishita, S. Fukushige, Y. Umeda, Y. (2015). Proposal of a computational design support method for sustainability scenario design. *Transactions of the JSME* (in Japanese), 81(822), p. 14-00269, http://doi.org/10.1299/transjsme.14-00269.
22. I.J. Petrick, A.E. Echols (2004). Technology roadmapping in review: A tool for making sustainable new product development decisions. *Technological Forecasting Social Change* 71, 81–100. https://doi.org/10.1016/S0040-1625(03)00064-7.
23. R. Fernandez-Flores, B. Hernandez-Morales, B. (2016). Towards an Open Learning Ecosystem Based on Open Access Repositories, Curricula-Based Indices and Learning Management Systems. In: L.G. Chova, A.L. Martinez,

I.C. Torres (eds.), *8th International Conference on Education and New Learning Technologies (EDULEARN)*, Barcelona, Spain, July 4–6, 2016—EDULEARN Proceedings, pp. 2826–2831.

24. J.J. Maldonado Mahauad, J.P. Carvallo, J.S. Zambrano (2016). Educational Repositories. Study of the Current Situation and Strategies to Improve Their Effective Use at Ecuadorian Universities. *IEEE Revista Iberoamericana De Tecnologias Del Aprendizaje* 11(2), 79–86. doi:10.1109/RITA.2016.2554001.

CHAPTER 10

A Robust Approach for Course of Action Comparison and Selection in Operation Planning Process

Ahmet Kandakoglu and Sarah Ben Amor

CONTENTS

10.1 INTRODUCTION

In the military, operation planning process (OPP) is a complex process that consists of logical steps to examine an assigned mission: develop, analyze, and compare alternative course of actions (COAs); select the best COA; and finally, produce a plan or order (JP 5-0, 2017). A COA can be defined as a potential solution to accomplish a mission. It is the main driver to develop an effective operation plan and orders at the end of the OPP.

Achieving success in an OPP depends first on developing and analyzing the COAs accurately and then comparing these COAs and selecting the best one for future development. Thus, comparing the alternative COAs and selecting the best one becomes a critical decision-making process within the OPP because this process yields better end-products. COA comparison is the process of evaluating all possible COAs and preparing a decision briefing, including a COA recommendation that will allow the commander to decide on the best one. In this process, COAs are not compared with each other but are rather evaluated independently with respect to a set of criteria (JP 5-0, 2017). There is no standard list of criteria because they differ from mission to mission. The commander and staff identify a list of important criteria that are sufficient to differentiate COAs for the assigned mission.

The key inputs of the COA comparison are the advantages and disadvantages, the wargaming results, the evaluation criteria, the estimates, and the preferences. The output of this process is the selected COA and the rationale behind this decision. Because of the time constraint and tremendous amount of information in this environment, the commander who is the actual decision maker cannot participate in all the activities during the decision-making process (Goztepe and Kahraman, 2015). The commander, upon receiving the staff's recommendation, should combine his or her analysis with the recommendation and make a final decision based on his or her experience and preferences.

Only limited or uncertain information is available to determine a solution to this complex decision-making problem because obtaining precise measurements and accurate preference information by an objective and comprehensive analysis of COAs is difficult and sometimes impossible. This is due to a number of reasons such as time constraints, disagreements among multiple groups, political sensitivity, and lack of commander involvement. For all these reasons, the commander is often unable or unwilling to express his or her preferences with precision and certainty in these situations. However, a decision has to be made based on the available information. Durbach and Calder (2016) called this kind of decision problem "low-involvement" decisions.

Therefore, in COA comparison and selection, the staff should not turn the decision into a mathematical equation or compel the commander to be more precise in his or her assessment of preference information. They should rather express clearly to the commander why one COA is preferred to another and provide him or her with additional information to support his or her final decision (JP 5-0, 2017).

Although COA is a crucial decision-making process in the military, there is a paucity of studies to support the COA selection process. A decision matrix based on the Weighted Sum Model (WSM) method is the most common technique in the official military documents (JP 5-0, 2017). Besides, WSM (Larsen and Herman, 1994), Analytic Hierarchy Process (AHP) (Minutolo, 2003), Simple Multi-Attribute Scoring Heuristic using Distance functions (SMASH-D) (Nunn, 2010), Value-Focused Thinking (VFT) (Ozdemir, 2013), and Multi-Attributive Border Approximation Area Comparison (MABAC) (Božanić, 2016) are the other methods applied. However, none of these studies deals with uncertain and incomplete data. To the best of our knowledge, only a multiple-criteria aggregation procedure called PAMSSEM (Guitouni et al., 1999; Bélanger and Guitouni, 2000) and the Fuzzy Analytic Network Process (FANP) (Goztepe et al., 2011) addressed uncertainties. PAMSSEM uses ELECTRE III type aggregation to construct outranking relations for qualitative, fuzzy, and probabilistic information and applies PROMETHEE II to rank the COAs based on these relations. In the FANP model, uncertainty is modeled as fuzzy linguistic variables in the pairwise comparison matrix. Nevertheless, these studies do not address the lack of commander involvement issue in the decision-making process and do not provide the staff with sufficient information to support their recommendation. Overall, there is still a requirement for MCDA approaches to fulfill these gaps in an effective and timely manner (Goztepe and Kahraman, 2015).

The SMAA-PROMETHEE has recently been proposed to investigate the robustness of multiple-criteria rankings when input parameters (evaluations and preference values) are uncertain or incomplete (Corrente et al., 2014). It addresses low-involvement decision making by providing information about the types of preferences that would lead to the selection of each alternative (Lahdelma et al., 1998; Lahdelma and Salminen, 2001; Durbach and Calder, 2016). It performs Monte-Carlo simulations and runs the PROMETHEE II (Brans and Vincke, 1985; Brans et al., 1986; Behzadian et al., 2010) as the MCDA method, which ensures that alternatives are assigned to specific ranks.

From this perspective, we propose a multi-criteria approach based on the SMAA-PROMETHEE method to help the staff and the commanders compare and select the COAs during the OPP. This approach also helps the commander to understand how the final ranking of COAs is sensitive to the different criteria weights and thresholds. It allows the commander to either select any COA by providing justifications or investigating the robustness of the ranking. Furthermore, this approach can provide significant time savings in the information collection stage by determining if the information is sufficiently accurate or needs to be refined.

The rest of this chapter is organized as follows. Section 10.2 overviews the SMAA-PROMETHEE method and its basic concepts. In Section 10.3, a description of the proposed approach is presented. Section 10.4 illustrates the approach through a case study. Conclusions and further studies are addressed, respectively, in the last section.

10.2 THE SMAA-PROMETHEE METHOD

PROMETHEE is a well-known family of outranking methods in MCDA developed by Brans and Vincke (1985) and further improved by Brans et al. (1986). It aggregates valued preference relations of every alternative pairs on each criterion to calculate the flow scores (entering, leaving, and net scores) and then provides a complete or partial ranking of alternatives based on these scores. Although PROMETHEE I builds a partial ranking of alternatives using the entering and leaving flows, PROMETHEE II provides a complete ranking of alternatives by considering the net flows. It is a rather simple and practical method that has been used in many real-world problems (Behzadian et al., 2010; Brans and De Smet, 2016). To apply the PROMETHEE method, the evaluations of the alternatives with respect to the criteria, the preference information for the criteria weights, indifference, and preference thresholds should be provided. Obviously, each set of these parameters may result in different rankings of alternatives.

The PROMETHEE method has been extended to incorporate uncertainties in the input parameters using various types of modeling. Fuzzy and interval-valued numbers-based approaches include those developed by Goumas and Lygerou (2000), Tuzkaya et al. (2010), Özgen et al. (2011), Chen et al. (2011), Senvar et al. (2014), Celik and Gumus (2016), Wu (2018), and Liang et al. (2018); another extension to deal with evidential evaluations (Abdennadher et al., 2013) and a framework to integrate mixed evaluations (stochastic, fuzzy, and evidential) (Ben Amor and Mareschal, 2012) have been proposed. In addition, Hyde et al. (2003) and Shakhsi-Niaei et al. (2011) developed stochastic approaches to PROMETHEE by embedding it into a Monte Carlo simulation framework. The output of simulation is used to obtain the probabilities of achieving different ranks by each alternative and also to analyze the impact of different uncertainties on the final ranking. However, the amount of information provided at the end of the simulation is insufficient for the decision makers to understand the preferences and rationale behind the rankings of the alternatives.

SMAA (Lahdelma et al., 1998; Lahdelma and Salminen, 2001) as a flexible and extendable framework can produce some possible stochastic values of the input parameters and apply inverse space analysis by running an MCDA method to describe the preferences that make each alternative the most preferred one. It simultaneously considers uncertainty in all parameters and quantifies all possible rankings for the alternatives in terms of probabilities. It also provides decision support through descriptive measures using a Monte Carlo simulation in practice. These advantages make SMAA applicable on a broad range of MCDA problems and, over the last decade, different SMAA variants based on traditional MCDA models were proposed to deal with various requirements (Tervonen and Figueira, 2008; Lahdelma and Salminen, 2010; Lahdelma and Salminen, 2016; Pelissari et al., 2019). Tervonen and Lahdelma (2007) presented the algorithms to implement SMAA methods through Monte Carlo simulation. Furthermore, Tervonen (2014) developed an open source JSMAA software that implements SMAA-2 and SMAA-TRI methods. Detailed information on different SMAA methods and applications can be found in the literature survey presented by Tervonen and Figueira (2008), and Pelissari et al. (2019).

The SMAA-PROMETHEE method has recently been proposed to investigate the robustness of multiple criteria rankings when input parameters for the PROMETHEE method are uncertain and incomplete (Corrente et al., 2014). In this combined method, SMAA generates possible parameter sets for the model, and the PROMETHEE method produces a ranking of alternatives for each of these sets.

Let us consider a decision problem represented as a set of m alternatives $\{x_1, x_2, \ldots, x_m\}$ that are evaluated in terms of n criteria. Uncertain or imprecise evaluation of x_i on criterion j is represented by stochastic variables ξ_{ij} with joint density function $f_X(\xi)$ in the space $X \sqsubseteq R^{m \times n}$. The weight space can be defined according to the requirements, but typically, the weights are non-negative and normalized. The weight space without any preference information is defined as follows:

$$W = \left\{ w \in R^n : w_j \geq 0, \forall j \, and \sum_{j=1}^{n} w_j = 1 \right\}$$

If there exists a preference information on weights, this space W is restricted by the constraints translated from this information. Imprecise thresholds are represented by the stochastic variables α_j and β_j corresponding to the deterministic thresholds p_j and q_j, respectively. To simplify the notation, we define these thresholds as $\tau = (\alpha, \beta)$ with joint density function f_τ in the space of possible threshold values.

SMAA analysis produces three different descriptive measures for decision support: the rank acceptability indices, the central weight vectors, and the pairwise winning indices. To compute these measures, let us define a ranking function that evaluates the rank r of the alternative x_i with the corresponding parameter values:

$$r(x_i) = rank(i, \xi, w, \tau)$$

This function executes PROMETHEE II method and returns the rank of an alternative based on its net flow.

The first and the main decision-aiding descriptive measure is the rank acceptability index b_i^r, which measures the variety of all different parameters that grant alternative x_i rank r. It is in the range [0, 1], and the best alternatives are those with higher acceptability indices for the best ranks. b_i^r is computed numerically as

$$b_i^r = \int_{W:rank(i,\xi,w,\tau)=r} f_w(w) \int_\chi f_\chi(\xi) \int_T f_T(\tau) dT d\xi dw$$

The central weight vector w_i^c is the expected centre of gravity of the favorable first-rank weights of an alternative. Presenting the central weights of different alternatives to the decision makers (DMs) may help them to understand how different weights correspond to different choices with the assumed preference model. w_i^c is computed using

$$w_i^c = \frac{1}{b_i^1} \int_\chi f_\chi(\xi) \int_T f_T(\tau) \int_{w \in W_i^1(\xi)} f_W(w) dw dT d\xi$$

The pairwise winning index p_{ik} is the probability of an alternative x_i being preferred to alternative x_k and can be computed as

$$p_{ik} = \int_{w \in W:rank(i,\xi,w,\tau) > rank(k,\xi,w,\tau)} f_w(w) \int_\chi f_\chi(\xi) \int_T f_T(\tau) dT d\xi dw$$

The pairwise winning indices are especially useful when trying to distinguish between the ranking differences of two alternatives that have similar rank acceptability indices (Leskinen et al., 2006; Tervonen et al., 2008). In these cases, looking at these indices between this pair of alternatives can help to determine whether or not one of them is superior to the other. In the SMAA-PROMETHEE simulation, the global pairwise winning index p_{ik} is obtained by comparing the alternatives x_i and x_k based on the net flows. Naturally, the simulation takes into account all the other alternatives. On the other hand, the local winning index is based on comparing the alternative pairs using the preference relations without taking into account all the others, and it is computed using the frequency of $\pi(x_i, x_k) > \pi(x_k, x_i)$, which represents how much alternative x_i is preferred to alternative x_k. For more information about the mathematical calculations, the readers may refer to the numerical example presented by Corrente et al. (2014).

10.3 DESCRIPTION OF THE PROPOSED APPROACH

The proposed approach is based on the SMAA-PROMETHEE method to address uncertainties in the COA comparison and selection process (Figure 10.1). The main inputs for this problem are the wargaming results, the staff estimates, the advantages and disadvantages of alternate COAs, and the criteria set that are all obtained in the COA analysis step. Using these inputs, an evaluation matrix is established with uncertainties expressed as a confidence interval, a statistical distribution, or an ordinal value. Wargaming outputs such as friendly aircraft lost or destroyed targets during the operation can be modeled as a normal distribution with a mean (μ) and standard deviation (σ). On the other hand, for some criteria such as the cost of the resources being used in COAs, evaluations can be any value within an interval. Finally, in some situations in which the staff can only have estimates on a criterion such as the possibility of failure, the staff may provide ordinal ranks of COAs with respect to this criterion.

The other inputs for the SMAA-PROMETHEE model are the preference information on the criteria weights and thresholds. The staff can provide some information on the indifference and preference thresholds in the form of interval values. They can also give different types of information regarding the weights of criteria, for example, a ranking of the criteria with respect to their importance, an interval of possible values for the criteria weights, different reference levels of the criteria, or a set of linear constraints. In extreme cases, especially in the early stages of the process, they

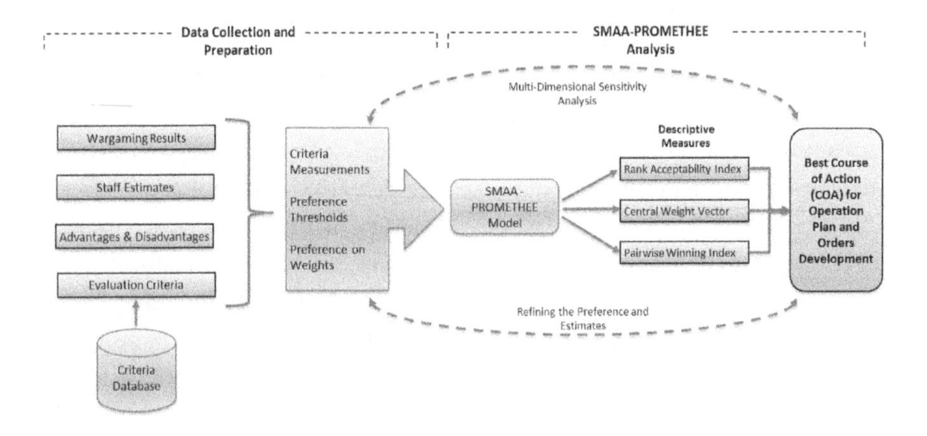

FIGURE 10.1 Schematic layout of the SMAA-PROMETHEE approach.

do not need to provide any preference information on weights because this approach can also provide practical solutions in this situation. Any additional information obtained at later stages can be included into the model as a linear constraint to increase the accuracy of the results.

The result of the SMAA-PROMETHEE model is obviously dependent on this information: the weights of criteria, the indifference and preference thresholds, and also the evaluations of the alternatives. Compatible samples of the model parameters are generated with the information provided, and for each sampled set of parameters, preference and indifference relations as well as the ranking of the alternatives, are obtained with the PROMETHEE II method. After performing a certain number of iterations, the rank acceptability indices, pairwise winning indices of alternatives, and the central weight vectors are computed as descriptive measures. Besides, the center of mass for criteria weights (the average preferences of the DM) and the average evaluations and thresholds for which each COA reached the first rank are the other outputs of the model.

The final output of this process is the ranking of the alternative COAs, the selected COA and the rationale behind the decision. Looking at all the outputs can provide valuable insights into the model results and the robustness of multi-criteria rankings. Furthermore, the staff can provide sufficient information to support their recommendation by addressing the lack of commander involvement in the decision-making process.

Based on the commander's feedback, this approach can be used in an interactive mode that makes it possible to iterate smoothly between presenting the recommended COA and refining the preference and evaluations.

In that way, the commander can understand the preferences that make each COA the most preferred one and also why the selected COA is preferred to others. Finally, this approach allows the staff and the commander to perform a multi-dimensional sensitivity analysis to test whether the results are dependent on the values assigned to the threshold preferences and weights.

10.4 CASE STUDY

In this section, a COA comparison and selection problem in the Air Operation Center is illustrated using the proposed SMAA-PROMETHEE approach. The staff developed five possible COAs and 10 evaluation criteria based on a given mission. The evaluation criteria are shown in Table 10.1. Friendly loss, destroyed targets, and collateral damage criteria are obtained from the simulation analysis and modeled as normal

TABLE 10.1 Evaluation Criteria

Criteria		Measurement	Polarity	Description
C1	Friendly Loss	Distribution	–	# of friendly aircraft lost during the operation
C2	Destroyed Targets	Distribution	+	# of enemy targets destroyed
C3	Collateral Damage	Distribution	–	# of civilian casualties caused by the collateral damage
C4	Operation Risk	Ordinal	–	The possibility of failure due to the unexpected events
C5	Surprise Effect	Ordinal	+	The unpredictability of the operation
C6	Operations Complexity	Ordinal	–	The CoA implementation difficulties caused by its operational requirements
C7	Command and Control Complexity	Ordinal	–	The CoA implementation difficulties caused by Command and control relationships and coordination requirements
C8	Logistics Complexity	Ordinal	–	The CoA implementation difficulties caused by its logistics requirements
C9	Sustainability	Ordinal	+	The ability to continue the operation during the time associated with the CoA
C10	Cost	Interval	–	The cost of the resources being used in million $

distributions. The cost criterion is given as an interval value with lower and upper bounds. The staff only provide an ordinal information for the operation risk, surprise effect, operations complexity, command and control complexity, logistics complexity, and sustainability criteria. An ordinal criterion is measured by ranking the alternatives according to that criterion. The evaluation of alternate COAs with respect to these criteria is presented in Tables 10.2 and 10.3.

A ranking of the criteria with respect to their importance, from the most to the least important, is provided for the criteria weights: *C*2, *C*1, *C*4, *C*3, *C*9, *C*5, *C*7, *C*6, *C*8, and *C*10. By transforming this information, the weight space is restricted by the following linear constraints:

$$w_{C2} \geq w_{C1} \geq w_{C4} \geq w_{C3} \geq w_{C9} \geq w_{C5} \geq w_{C7} \geq w_{C6} \geq w_{C8} \geq w_{C10}$$

Regarding the indifference and the preference thresholds, the staff assigned the values presented in Table 10.4. For the sake of simplicity, a linear preference function is used in this study to represent the commander's preference.

TABLE 10.2 COA Information for Distributional and Interval Criteria

	C1	C2	C3	C10
COA-1	$\mu = 21, \sigma = 3$	$\mu = 95, \sigma = 5$	$\mu = 15, \sigma = 2$	[20, 30]
COA-2	$\mu = 9, \sigma = 2$	$\mu = 79, \sigma = 6$	$\mu = 14, \sigma = 2$	[15, 20]
COA-3	$\mu = 24, \sigma = 3$	$\mu = 98, \sigma = 2$	$\mu = 12, \sigma = 3$	[35, 40]
COA-4	$\mu = 32, \sigma = 5$	$\mu = 86, \sigma = 4$	$\mu = 19, \sigma = 4$	[20, 25]
COA-5	$\mu = 12, \sigma = 2$	$\mu = 83, \sigma = 5$	$\mu = 10, \sigma = 2$	[10, 15]

TABLE 10.3 COA Information for Ordinal Criteria

	C4	C5	C6	C7	C8	C9
COA-1	4	4	4	3	2	3
COA-2	1	1	2	1	4	5
COA-3	2	5	5	3	3	4
COA-4	5	2	1	5	1	1
COA-5	3	3	3	4	5	2

TABLE 10.4 Indifference and Preference Thresholds

	C1	C2	C3	C4	C5	C6	C7	C8	C9	C10
Q	[0, 3]	[1, 3]	[1, 3]	[0, 0.03]	[0, 0.03]	[0, 0.03]	[0, 0.03]	[0, 0.03]	[0, 0.03]	[0, 3]
P	[3, 5]	[5, 10]	[5, 10]	[0.05, 0.1]	[0.05, 0.1]	[0.05, 0.1]	[0.05, 0.1]	[0.05, 0.1]	[0.05, 0.1]	[3, 5]

We performed 10,000 Monte Carlo iterations for SMAA computations with the sampled sets of parameters compatible with the above information. The three descriptive measures, the rank acceptability indices, the central weight vectors, and the pairwise winning indices as well as the center of mass for criteria weights are provided in Tables 10.5 through 10.8 to support the commander's decision. According to rank acceptability indices presented in percents in Table 10.5, COA-5 seems to be the best COA with a 44% probability to obtain the first rank. On the other hand,

TABLE 10.5 Rank Acceptability Indices (in percentage)

	Rank 1	Rank 2	Rank 3	Rank 4	Rank 5
COA-1	42	34	16	6	01
COA-2	01	2	4	14	80
COA-3	08	21	37	30	03
COA-4	05	12	28	41	15
COA-5	44	32	15	8	01

TABLE 10.6 Pairwise Winning Indices (in percentage)

	COA-1	COA-2	COA-3	COA-4	COA-5
(a) Global Pairwise Winning Indices					
COA-1	0	98	79	84	49
COA-2	02	0	06	16	03
COA-3	21	94	0	63	23
COA-4	16	84	37	0	15
COA-5	51	97	74	85	0
(b) Local Pairwise Winning Indices					
COA-1	0	98	80	80	37
COA-2	02	0	1	31	09
COA-3	19	99	0	88	06
COA-4	20	68	12	0	59
COA-5	63	90	93	41	0

TABLE 10.7 Central Weight Vectors

	C1	C2	C3	C4	C5	C6	C7	C8	C9	C10
COA-1	0.195	0.316	0.105	0.009	0.143	0.060	0.031	0.044	0.019	0.080
COA-2	0.231	0.320	0.098	0.008	0.132	0.055	0.026	0.038	0.018	0.074
COA-3	0.184	0.388	0.092	0.008	0.122	0.052	0.029	0.039	0.018	0.069
COA-4	0.181	0.275	0.116	0.009	0.148	0.074	0.032	0.052	0.019	0.094
COA-5	0.193	0.253	0.117	0.012	0.146	0.071	0.038	0.053	0.024	0.092

TABLE 10.8 Central of Mass

C1	C2	C3	C4	C5	C6	C7	C8	C9	C10
0.193	0.292	0.110	0.010	0.143	0.065	0.034	0.048	0.021	0.085

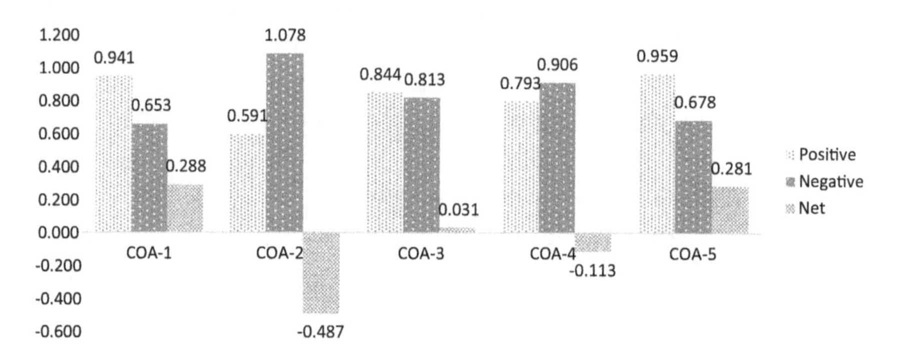

FIGURE 10.2 Average of flows.

COA-1 has a very close first rank probability (42%), which makes it difficult for the commander to distinguish the ranking differences to make a choice between these two COAs. The average net flow of COA-1 is slightly higher than the one for COA-5, which makes the choice even more difficult (Figure 10.2).

At this point, it is useful to examine the global pairwise winning indices which show how much an alternative is preferred to all other alternatives. COA-5 is preferred to all the other COAs with a frequency of at least 51% (the fifth row of Table 10.6a). Actually, the value 51% shows that again the choice between COA-5 and COA-1 is difficult, as COA-1 is slightly preferred to COA-5 with a frequency of 49%. The global winning index is based on the net flows and takes into account all the other alternatives. On the other hand, the local winning index is based on the relation matrix in PROMETHEE II and compares the alternative pairs without taking into account all the others. The frequency of 63% in Table 10.6b clearly points out that COA-5 is preferred to COA-1 more than COA-5 is preferred to COA-1. Based on these findings, the commander should be more comfortable that COA-5 is the best COA in this situation.

COA-2 is the worst one because it has lower acceptability indices for the best ranks in Table 10.5. In addition, the second column of Table 10.6a shows that each COA is preferred to COA-2 with a frequency of at least 84%. COA-4 is the second worst option because of the indices in Tables 10.5 and 10.6. As a result, the final ranking of the alternatives is COA-5, COA-1, COA-3, COA-4, and COA-2.

It is observed that every COA could achieve the first position (Table 10.5), and therefore, it is important to determine which parameter values make this happen. The central weight vectors presented in percentages in Table 10.7, and the average evaluations and thresholds for which each COA reaches the first rank in Table 10.9, provide valuable insights into this situation. To make recommendations to the commander, we need to identify the strengths and weaknesses of the evaluated COA's based on this information.

Generally, if the central weight for a COA is high, then the COA is performing better than the remaining COAs with respect to that benefit criterion or vice-versa. For instance, COA-2 has the minimum number of friendly aircraft losses among the others, and this is the second-most important criterion (C1). Thus, although performing the worst in the most important criterion, a higher value of 0.231 for the central weight of C1 makes the COA-2 the best in some situations.

Besides, the staff may be interested in understanding the robustness of the final results. For this purpose, we performed a sensitivity analysis by applying the same procedure as in Corrente et al. (2014) to test whether the results provided by the approach are dependent on the values assigned to the threshold preferences and weights. First, we randomly generated 2000 problems by sampling an evaluation table, M, and a vector of weights, w, compatible with the information provided by the staff and the commander. Then, for each problem (set of M and w), we sampled 10,000 thresholds in the considered intervals. We studied the rank acceptability indices obtained by applying the proposed approach.

Looking at the second column in Table 10.10, we could state that, fixing the evaluation table and the weight vector, in some cases the variation of the thresholds does not influence the results of the model, whereas in all others cases it could marginally influence the results. For instance, in 1409 problems, only one COA ranked first when the thresholds were changed. Furthermore, only in 1 problem all COAs could have the first rank, whereas in 2 problems only four out of the five COAs could fill the first position. This means that, in this case study, once the evaluation table and the criteria weights have been fixed, the variation of the thresholds would not highly influence the results obtained by using the approach.

TABLE 10.9 Average Evaluations and Thresholds for Which Each COA Reached the First Rank

	C1	C2	C3	C4	C5	C6	C7	C8	C9	C10
(a) COA-1										
COA-1	20.37	97.50	14.63	24.96	0.24	0.25	0.24	0.49	0.74	0.50
COA-2	8.92	79.48	13.99	17.51	1.00	1.00	0.75	1.00	0.24	0.00
COA-3	24.41	97.83	12.07	37.48	0.75	0.00	0.00	0.49	0.49	0.25
COA-4	32.21	86.03	19.14	22.51	0.00	0.75	1.00	0.00	1.00	1.00
COA-5	12.17	81.50	10.24	12.47	0.50	0.50	0.50	0.25	0.00	0.75
q_j	1.50	2.01	1.99	1.48	0.01	0.02	0.02	0.02	0.01	0.02
p_j	4.00	7.46	7.57	3.99	0.07	0.08	0.07	0.07	0.08	0.07
(b) COA-2										
COA-1	21.73	89.76	14.75	25.55	0.25	0.19	0.24	0.52	0.76	0.51
COA-2	8.16	92.98	13.27	17.50	1.00	1.00	0.70	1.00	0.27	0.00
COA-3	24.68	97.44	11.98	37.58	0.78	0.00	0.00	0.52	0.50	0.20
COA-4	32.45	85.18	18.91	22.43	0.00	0.73	1.00	0.00	1.00	1.00
COA-5	12.44	80.53	9.59	12.57	0.50	0.46	0.49	0.28	0.00	0.80
q_j	1.55	2.01	1.99	1.33	0.02	0.02	0.02	0.01	0.01	0.02
p_j	3.86	7.61	7.51	4.17	0.08	0.07	0.07	0.07	0.08	0.07
(c) COA-3										
COA-1	21.94	90.92	15.33	25.06	0.28	0.23	0.24	0.50	0.76	0.50
COA-2	8.89	80.18	13.94	17.55	1.00	1.00	0.75	1.00	0.25	0.00
COA-3	22.82	98.75	10.88	37.59	0.74	0.00	0.00	0.50	0.50	0.26
COA-4	32.41	86.32	19.01	22.58	0.00	0.74	1.00	0.00	1.00	1.00
COA-5	12.21	81.48	10.13	12.53	0.52	0.49	0.49	0.26	0.00	0.75
q_j	1.48	1.95	1.99	1.48	0.01	0.02	0.01	0.02	0.02	0.01
p_j	3.99	7.35	7.46	4.01	0.08	0.07	0.07	0.08	0.08	0.07
(d) COA-4										
COA-1	21.62	91.59	15.17	24.97	0.27	0.25	0.26	0.51	0.75	0.48
COA-2	8.67	79.21	13.92	17.50	1.00	1.00	0.76	1.00	0.26	0.00
COA-3	24.40	97.86	12.02	37.44	0.75	0.00	0.00	0.51	0.51	0.23
COA-4	29.83	90.86	16.04	22.49	0.00	0.76	1.00	0.00	1.00	1.00
COA-5	12.08	79.77	10.41	12.58	0.53	0.50	0.51	0.26	0.00	0.75
q_j	1.51	2.00	2.04	1.50	0.01	0.01	0.01	0.02	0.01	0.02
p_j	3.97	7.48	7.64	3.95	0.07	0.08	0.07	0.07	0.07	0.08
(e) COA-5										
COA-1	21.32	93.71	15.27	24.97	0.25	0.24	0.25	0.50	0.76	0.49
COA-2	9.13	78.01	14.00	17.49	1.00	1.00	0.75	1.00	0.25	0.00
COA-3	23.82	98.11	12.06	37.48	0.75	0.00	0.00	0.50	0.50	0.25
COA-4	31.92	85.51	19.02	22.48	0.00	0.75	1.00	0.00	1.00	1.00
COA-5	11.84	85.29	9.69	12.50	0.49	0.50	0.50	0.24	0.00	0.75
q_j	1.52	2.01	1.99	1.52	0.01	0.02	0.02	0.02	0.02	0.02
p_j	3.99	7.52	7.44	4.01	0.08	0.07	0.08	0.08	0.07	0.08

TABLE 10.10 Robustness Analysis on Thresholds and Weights

Number of COAs Having First Rank	Number of Problems (on thresholds)	Number of Problems (on weights)
1	1409	123
2	527	1128
3	61	571
4	2	171
5	1	7

Similarly, we tested whether the results provided by our approach are dependent on the values assigned to the criteria weight preferences. First, we randomly generated 2000 problems by sampling an evaluation table *M* and vectors of thresholds *t*, and then for each problem (set of *M* and *t*), we randomly generated 10000 weights compatible with the order of importance provided before we finally applied the approach.

Based on the results at the third column of Table 10.10, we may conclude that, fixing the evaluation table and the thresholds, the variation of the weights could marginally influence the results of the model. For instance, in 123 problems, only one COA ranked first when the weights where changed. Additionally, in 1128 problems, two COAs could have the first rank, whereas in 571 problems three out of the five COAs could fill the first position. This means that, in this case study, the variation of the weights would highly influence the results once the evaluation table and the thresholds have been fixed.

10.5 CONCLUSION

The combined SMAA-PROMETHEE approach proposed in this study provides a scientific basis to make robust decisions based on uncertain and incomplete information for the COA comparison and selection. It also allows the staff to support the commander's decision-making process by clearly portraying the commander's options with high-dimensional sensitivity analysis, which has the flexibility to change all possible model parameters (criteria weights and preference thresholds) at the same time. In other words, it addresses this low-involvement decision-making problem by providing information about the input parameters that would lead to the selection of each alternative.

The nature of the problem shows that especially in the early stages of the decision-making process, the inputs to the preference model and evaluations either cannot be assessed at all or can only be assessed within relatively large bounds of uncertainty. The SMAA-PROMETHEE approach appears as a suitable methodology to save costs and time in the data elicitation process. For instance, even in such an extreme case where no criteria weight information is available, it is possible to provide practical solutions.

Based on the fact that the staff has to deal with a large volume of information within a very short time period during an operation, a decision support system based on the proposed approach would be helpful. It makes it possible to perform high-dimensional sensitivity and robustness analysis of the COA comparison scenarios in a timely manner. As a further research, a DSS should be developed and integrated into the current command and control systems. Furthermore, this approach can be extended to other large-scale military problems that include incomplete and uncertain information and can cope with the issue of low involvement of commanders in the decision process.

REFERENCES

Abdennadher, H., Boujelben, M. A., & Ben Amor, S. (2013). An extension of PROMETHEE with Evidential Evaluations. *Journal of Information Technology Review*, 4(3), 115–125.

Behzadian, M., Kazemzadeh, R. B., Albadvi, A., & Aghdasi, M. (2010). PROMETHEE: A comprehensive literature review on methodologies and applications. *European Journal of Operational Research*, 200(1), 198–215.

Bélanger, M., & Guitouni, A. (2000). A decision support system for CoA selection. In *Proceedings of the 5th International Command and Control Research and Technology Symposium*, Canberra ACT, Australia (www.dodccrp.org/events/5th_ICCRTS/papers/Track5/049.pdf).

Ben Amor, S., & Mareschal, B. (2012). Integrating imperfection of information into the PROMETHEE multicriteria decision aid methods: A general framework. *Foundations of Computing and Decision Sciences*, 37(1), 9–23.

Božanić, D. I., Pamučar, D. S., & Karović, S. M. (2016). Application the MABAC method in support of decision-making on the use of force in a defensive operation. *Tehnika*, 71(1), 129–136.

Brans, J. P., & De Smet, Y. (2016). PROMETHEE methods. In S. Greco, M. Ehrgott, J. Figueira (eds.) *Multiple Criteria Decision Analysis* (pp. 187–219). Springer, New York.

Brans, J. P., & Mareschal, B. (2005). PROMETHEE methods. In J. Figueira, S. Greco, & M. Ehrgott (eds.), *Multiple Criteria Decision Analysis: State of the Art Surveys* (pp. 163–196). Springer, Berlin, Germany.

Brans, J. P., & Vincke, P. (1985). Note—A preference ranking organisation method: (The PROMETHEE Method for Multiple Criteria Decision-Making). *Management Science*, 31(6), 647–656.

Brans, J. P., Vincke, P., & Mareschal, B. (1986). How to select and how to rank projects: The PROMETHEE method. *European Journal of Operational Research*, 24(2), 228–238.

Celik, E., & Gumus, A. T. (2016). An outranking approach based on interval type-2 fuzzy sets to evaluate preparedness and response ability of non-governmental humanitarian relief organizations. *Computers & Industrial Engineering*, 101, 21–34.

Chen, Y. H., Wang, T. C., & Wu, C. Y. (2011). Strategic decisions using the fuzzy PROMETHEE for IS outsourcing. *Expert Systems with Applications*, 38(10), 13216–13222.

Corrente, S., Figueira, J., & Greco, S. (2014). The SMAA-PROMETHEE method, *European Journal of Operational Research*, 239(2), 514–522.

Durbach, I. N., & Calder, J. M. (2016). Modelling uncertainty in stochastic multi-criteria acceptability analysis. *Omega*, 64, 13–23.

Goumas, M., & Lygerou, V. (2000). An extension of the PROMETHEE method for decision making in fuzzy environment: Ranking of alternative energy exploitation projects. *European Journal of Operational Research*, 123(3), 606–613.

Goztepe, K., Ejder, A., & ve Calikoglu, E. (2011). Course of Action (COA) Selection for special operations using Fuzzy multi-criteria decision making technique. *Proceedings of the Second International Fuzzy Systems Symposium* (pp. 355–359), November 17–18, 2011, Ankara, Turkey.

Goztepe, K., & Kahraman, C. (2015). A new approach to military decision making process: Suggestions from MCDM point of view. In S. Cetin and K. Goztepe (eds.) *International Conference on Military and Security Studies*, İstanbul, Turkey (pp. 118–122).

Guitouni, A., Bélanger, M., & Martel, J. M. (1999). A multiple criteria aggregation procedure for the evaluation of courses of action in the context of the Canadian Airspace Protection. *Defence Research Establishment Valcartier.* DREV-TR-1999-215.

Hyde, K., Maier, H. R., & Colby, C. (2003). Incorporating uncertainty in the PROMETHEE MCDA method. *Journal of Multi-Criteria Decision Analysis*, 12(4–5), 245–259.

Joint Publication (JP) 5-0. (2017). *Joint Planning. Joint Chiefs of Staff*, Washington, DC, June 16, 2017 (https://www.jcs.mil/Portals/36/Documents/Doctrine/pubs/jp5_0_20171606.pdf).

Lahdelma, R., Hokkanen, J., & Salminen, P. (1998). SMAA: Stochastic multiobjective acceptability analysis. *European Journal of Operational Research* 106, 137–143.

Lahdelma, R., Salminen, P. (2001). SMAA-2: Stochastic multicriteria acceptability analysis for group decision making. *Operations Research*, 49(3), 444–454.

Lahdelma, R., & Salminen, P. (2010). Stochastic multicriteria acceptability analysis (SMAA). In Greco, S., Ehrgott, M., & Figueira, J. R. (Eds.). *Trends in Multiple Criteria Decision Analysis* (pp. 285–315). Springer, Boston, MA.

Lahdelma, R., & Salminen, P. (2016). SMAA in robustness analysis. In *Robustness Analysis in Decision Aiding, Optimization, and Analytics* (pp. 1–20). Springer International Publishing, Basel, Switzerland.

Larsen, R. W., & Herman, J. S. (1994). Course-of-action selection tool COAST. In *Knowledge-Based Planning and Scheduling Initiative: Workshop Proceedings*. February 21–24, 1994 (p. 235). Morgan Kaufmann, Tucson, Arizona.

Leskinen, P., Viitanen, J., Kangas, A., & Kangas, J. (2006). Alternatives to incorporate uncertainty and risk attitude in multicriteria evaluation of forest plans. *Forest Science*, 52(3), 304–312.

Liang, R. X., Wang, J. Q., & Zhang, H. Y. (2018). Projection-based PROMETHEE methods based on hesitant fuzzy linguistic term sets. *International Journal of Fuzzy Systems*, 20(7), 2161–2174.

Minutolo, M. (2003). Use of analytic hierarchy process modeling in the military decision making process for course of action evaluation and unit cohesion. In *International Symposium of Analytic Hierarchic Process* (pp. 347–348), Bali, Indonesia.

Nunn, L. R. (2010). Enhancing the military decision making process with a simple multi-attribute scoring heuristic using distance functions (SMASH-D), MSc thesis, The University of Texas at Austin, Austin, TX.

Ozdemir, A. (2013). Evaluating courses of actions at the strategic planning level (No. AFIT-ENS-13-M-14). Air Force Institute of Technology, Graduate School of Engineering and Management, OH.

Özgen, A., Tuzkaya, G., Tuzkaya, U. R., & Özgen, D. (2011). A multi-criteria decision making approach for machine tool selection problem in a fuzzy environment. *International Journal of Computational Intelligence Systems*, 4(4), 431–445.

Pelissari, R., Oliveira, M. C., Ben Amor, S., & Abackerli, A. J. 2019. A new FlowSort-based method to deal with information imperfections in sorting decision-making problems. *European Journal of Operational Research*. https://doi.org/10.1016/j.ejor.2019.01.006.

Pelissari, R., Oliveira, M. C., Ben Amor, S., Kandakoglu A., Helleno A. L. 2019. SMAA methods and their applications: A literature review and future research directions. *Annals of Operations Research*. https://doi.org/10.1007/s10479-019-03151-z.

Senvar, O., Tuzkaya, G., & Kahraman, C. (2014). Multi criteria supplier selection using fuzzy PROMETHEE method. In *Supply Chain Management under Fuzziness* (pp. 21–34). Springer, Berlin, Germany.

Shakhsi-Niaei, M., Torabi, S. A., & Iranmanesh, S. H. (2011). A comprehensive framework for project selection problem under uncertainty and real-world constraints. *Computers & Industrial Engineering*, 61(1), 226–237.

Tervonen, T. (2014). JSMAA: Open source software for SMAA computations. *International Journal of Systems Science*, 45(1), 69–81. https://doi.org/10.1080/00207721.2012.659706.

Tervonen, T., & Figueira, J. R. (2008). A survey on stochastic multicriteria acceptability analysis methods. *Journal of Multi-Criteria Decision Analysis*, 15, 1–14.

Tervonen, T., Figueira, J. R., Lahdelma, R., & Salminen, P. (2008). SMAA-III: A simulation-based approach for sensitivity analysis of ELECTRE III. In *Real-Time and Deliberative Decision Making* (pp. 241–253). NATO Science for Peace and Security Series C: Environmental Security. Springer, Dordrecht, the Netherlands.

Tervonen, T., & Lahdelma, R., (2007). Implementing stochastic multicriteria acceptability analysis. *European Journal of Operational Research*, 178(2), 500–513.

Tuzkaya, G., Gülsün, B., Kahraman, C., & Özgen, D. (2010). An integrated fuzzy multi-criteria decision making methodology for material handling equipment selection problem and an application. *Expert Systems with Applications*, 37(4), 2853–2863.

Wu, Y., Xu, C., Ke, Y., Chen, K., & Sun, X. (2018). An intuitionistic fuzzy multi-criteria framework for large-scale rooftop PV project portfolio selection: Case study in Zhejiang, China. *Energy*, 143, 295–309.

CHAPTER 11

Analyzing the Relationship between Human Development and Competitiveness Using DEA and Cluster Analysis

Hakan Kılıç and Özgür Kabak

CONTENTS

11.1 INTRODUCTION

Philosophers, economists, and political leaders have been emphasizing that the main purpose of development is to reach human welfare (Edewor, 2014). References in line with this claim can be found at the manuscripts of Aristotle, Emmanuel Kant, Adam Smith, Karl Marx, John Stuart Mill, and many more philosophers and political economists (Jiyad, 1998). Human development, in this regard, investigates the basic development idea, namely human welfare.

Ülengin et al. (2011) states that if the competitiveness of a country is managed properly, then the level of human well-being is expected to be one of the key results. It is necessary for the countries to use their competitiveness to advance their social welfare, or in other words, their human development. On the other hand, when a country advances in their human development, it is likely that their competitiveness level also increases.

According to Seers (1972) development is an opportunity to create conditions to realize the human potential. Also, Anand & Sen (1992) state that people should be placed at the center of the development to enable equal opportunities for all regarding freedom, achievements, and skill. As such, human development covers all aspects of human life such as social, economic, cultural, and political (Ciko, 2015). On the other hand, United Nations Development Programme (UNDP) (1990) originally defined human development as the process enlarging people's choices and included healthy life, knowledge acquisition, and achievement of decent living standards, and developed the Human Development Index (HDI) as a measurement tool. This index was established thanks to Amartya Sen's (Sen, 1985, 1992) capability approach (Bérenger & Verdier-Chouchane, 2007). It comprises the dimensions of health, education, and income. Even though HDI has its own shortcomings, it is the widely used indicator of human development in the literature (Buscema et al., 2016).

Competitiveness at county level is first defined by Porter (1990): It is the innovation capacity of a country, which allows the country to be in a better position or to keep its favorable position in the key sectors in comparison to the other countries. As such Porter (1990) dismisses labor cost, exchange rate economies of scale or vast natural resources, and instead established productivity as the competitiveness of a country (Thore & Tarverdyan, 2016). World Economic Forum (WEF) (2011) defines the competitiveness as the government's politics, regulations, and legislations, which enables sustainable economic development and long-term prosperity. Additionally, WEF has developed the Global Competitiveness Index (GCI) to measure the competitiveness. It consists of 12 pillars: institutions, infrastructure, macroeconomic environment, health and primary education, higher education and training, goods market efficiency, labor market efficiency, financial market development, technological readiness, market size, business sophistication, and innovation. Among the competitiveness indicators GCI is the most valued and widely accepted (Dudas, 2014; Perez-Moreno et al., 2016).

There are limited studies in the literature investigating the human development and competitiveness relationship. Lonska and Boronenko (2015) investigated the relationship between human development and national competitiveness and showed that GCI and HDI growth trends are positively correlated. Ülengin et al. (2011) examined with which level of efficiency the countries managed to transform their competitiveness to human development, the main aim of human activities, via DEA and used these results to identify the main factors affecting the efficiency by artificial neural network methods. Thore and Tarverdyan (2016) have examined the level of countries' efficiencies regarding the conversion of their sustainable competitiveness into environmental and social welfare by using DEA. To the best of our knowledge, there is no study taking into account the interrelations and bilateral relations among human development and competitiveness drivers with time lags.

This study claims that there is bilateral relation between human development and competitiveness. We analyze this relation at country level using data envelopment analysis (DEA) and cluster analysis. To that end, output-oriented variable returns to scale DEA model with time lags is used. Window analysis and cluster analysis is incorporated to this model to observe the trend. Furthermore, the results are discussed to understand the direction of the effect in the human development and competitiveness relationship. In fact, DEA is not a methodology for exploring a causal relationship. DEA was originally developed for performance measurement and relative efficiency of decision-making units (DMU) based on their input and output levels. In a DEA setting, it is explicitly assumed that there is input-output relation between specified factors, that is, in general, smaller levels of inputs and larger levels of outputs represent better efficiency (Cook et al., 2014). In this study, it is not clear which one, human development and competitiveness, is the input or the output. Therefore, we developed two DEA models: DEA-1, human development is the input, competitiveness is the output; DEA-2, competitiveness is the input, human development is the output. We calculate efficiencies of the countries with both models. The results are compared to find out which DEA model is consistent. The input-output definition in the more consistent model shows the direction of the effect in the human development and competitiveness relationship.

As data sources, GCI of WEF's Global Competitiveness Report and HDI of UNDP's Human Development Report are used. GCI compares the countries in terms of policies and factors that determine the level of economic productivity. It is composed of three subindexes, namely, basic requirements, efficiency enhancers, and innovation and sophistication factors. HDI was created with the idea that the ultimate criteria for assessing the development of a country should be people and their capabilities, not economic growth alone. It is a summary measure of average achievement in key dimensions of human development: a long and healthy life, being knowledgeable, and have a decent standard of living. As presented in Figure 11.1, the subindexes of GCI and key dimensions of HDI are taken into account in our analysis.

After analyzing the relations between human development and competitiveness, we use the results of DEA and cluster analysis that incorporate the subindexes of GCI and dimensions of HDI to evaluate the

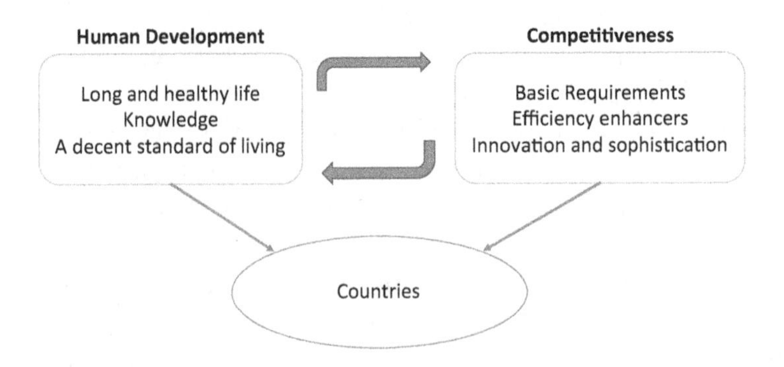

FIGURE 11.1 Framework of the study.

countries. In this respect, the problem can be considered as a multi-criteria decision-making (MCDM) problem, where the countries are alternatives and three subindexes of GCI, three dimensions of HDI, and DEA scores are the criteria. Although DEA is not a conventional MCDM method, it is an appropriate tool to approach MCDM problems (Bouyssou, 1999). The DMU in DEA correspond to alternatives in MCDM, and inputs and outputs correspond to criteria. The efficiency in DEA is related to convex efficiency in MCDM (Bouyssou, 1999). There are also some difficulties in the usage of DEA in a MCDM problem. A serious difficulty that may also be valid in our application is that efficient alternatives in DEA results may not be convex efficient (i.e., a country with low scores in all HDI and GCI indices may be found as efficient in DEA). To overcome this issue, we use cluster analysis to distinguish the efficient countries with low scores from the ones with high scores.

The rest of the chapter is structured as follows. Section 11.2 reviews the literature on competitiveness and human development relationship and DEA. Section 11.3 presents the methodology. Empirical results are given in Section 11.4, and Section 11.5 concludes the study.

11.2 LITERATURE REVIEW

In this review, we first present the studies on competitiveness and human development interaction. Subsequently, we explained DEA in details.

11.2.1 Competitiveness and Human Development

The literature regarding the relationship between human development and competitiveness is limited. Shkiotov (2013) presented the world's five-largest economies for a 5-year period via tables in terms of national competitiveness, life, and productivity levels by using GCI and HDI and compared them without using mathematical methods; and contrary to M. Porter's notion of dependency among productivity, quality of life, and national competitiveness, he has come to the conclusion that there is no direct relationship between these dimensions. Skorvagova and Drienikova (2016) claim that there is no relation between the rankings of European Union member states' national competitiveness and their welfare state type classes according to Esping-Andersen's definitions.

Conversely, Waheeduzzaman (2002) admitting the ultimate goal of competitiveness is to ensure the prosperity of the citizens; examined the impact of national competitiveness on per-capita income, human development, and inequity, and as a result, he proved via correlation analysis that competitiveness has positive impact on human development. Additionally, Onyusheva (2015) has investigated the significance of human capital in regard to the development of national economic competitiveness via correlation analysis between GCI and HDI for 144 countries on the time period 2012–2013 and concluded that a relationship exists. Furthermore, Lonska and Boronenko (2015) also investigated the relationship between human development and national competitiveness and showed that GCI and HDI growth trends are positively correlated, and similar to the literature on countries with GCI growth trends, they have identified that countries with high HDI scores having low capacity for further advancements in human development. Similarly, Bucher (2018) observed high correlation between a country's competitiveness and human development.

Moreover, Aiginger (1998) has provided a framework for measuring the competitiveness of countries by associating competitiveness with the notion of maximizing welfare level. In addition, Aiginger (2006) suggested that competitiveness should be defined as a level of welfare for a country or a particular region and, accordingly, presented a model for measuring competitiveness via its output, welfare. On the other hand, Ülengin et al. (2009) investigated the relationship between competitiveness and human development of countries via DEA and categorized the countries in light of these results by clustering. In addition, Ülengin et al. (2011) examined with which level of efficiency the countries managed to transform their competitiveness to human development, the main aim of human activities, via DEA and used these results to identify the main factors affecting the efficiency via artificial neural network methods. Ditkun et al. (2014) examined Brazil's competitiveness in the globe for especially 2003–2013 time period and arrived at the conclusion that the country has lost its competitiveness, and argued that this may be the case because of its fixed HDI score since 2007. Also, Thore and Tarverdyan (2016) have examined the level of countries' efficiencies regarding the conversion of their sustainable competitiveness into environmental and social welfare by using DEA. On the other hand, Tridico & Meloni (2018) used econometric analysis and observed that threats to competitiveness can be handled better by increasing human development investments.

11.2.2 Data Envelopment Analysis

DEA is a widely used nonparametric data-oriented approach to evaluate the performance of a set of peer entities. There is a rapid increase in DEA-related articles (Emrouznejad & Yang, 2018). Liu et al. (2013) investigated the DEA literature from 1978 to 2010 and identified that banking, health care, agriculture and farming, transportation, and education are the top five DEA application areas, which comprise approximately 41% of total papers. Additionally; agriculture, banking, supply chain, transportation, and public policy are the top five application fields of DEA in 2015 and 2016 (Emrouznejad & Yang, 2018). On the other hand, Liu et al. (2016) state that bootstrapping and two-stage analysis, undesirable factors, cross-efficiency and ranking, and network DEA, dynamic DEA, and slacks-based measure are the recent research fronts in DEA.

The basic DEA models in the literature are the constant returns to scale (Charnes et al., 1978) and variable returns to scale (Banker et al., 1984). Malmquist index (Färe et al., 1994) and window analysis (Charnes et al. 1985) are the unique extensions of DEA model to analyze longitudinal data (Sánchez, 2018). These four models are presented in this subsection.

11.2.2.1 Constant Returns to Scale

Suppose that there are N DMU. Each DMU^i is the ith DMU ($i = 1, \ldots, N$) with a s-dimensional input vector $\boldsymbol{x}^i = (x_1^i, \ldots, x_s^i)'$ and a r-dimensional output vector $\boldsymbol{y}^i = (y_1^i, \ldots, y_r^i)'$. Then, the following input and output matrices are formed:

$$\boldsymbol{X} = \left[\boldsymbol{x}^1, \ldots, \boldsymbol{x}^N\right]$$

$$\boldsymbol{Y} = \left[\boldsymbol{y}^1, \ldots, \boldsymbol{y}^N\right]$$

The input-oriented constant returns to scale (CRS) model in envelopment form is the following (Charnes et al., 1978):

$$\begin{aligned}
\text{minimize} \quad & \theta - \varepsilon \mathbf{1}'\left(\boldsymbol{s}^- + \boldsymbol{s}^+\right) \\
\text{subject to} \quad & \boldsymbol{X}\Lambda + \boldsymbol{s}^- = \theta \boldsymbol{x}^i \\
& \boldsymbol{Y}\Lambda - \boldsymbol{s}^+ = \boldsymbol{y}^i \\
& \boldsymbol{s}^-, \boldsymbol{s}^+, \Lambda \geq 0
\end{aligned}$$

where $\boldsymbol{s}^-$ and $\boldsymbol{s}^+$ are slack variable vectors and $\varepsilon > 0$ is a non-Archimedean element defined to be smaller than any positive real number. This model is equivalent to first solving the following:

$$\theta^i = \text{minimize} \quad \theta$$

$$\text{subject to} \quad \boldsymbol{X\Lambda} + \boldsymbol{s}^- = \theta \boldsymbol{x}^i$$

$$\boldsymbol{Y\Lambda} - \boldsymbol{s}^+ = \boldsymbol{y}^i$$

$$\boldsymbol{s}^-, \boldsymbol{s}^+, \boldsymbol{\Lambda} \geq 0$$

and then solving the below model by fixing the value of θ with the found solution:

$$\text{maximize} \quad \mathbf{1}'\left(\boldsymbol{s}^- + \boldsymbol{s}^+\right)$$

$$\text{subject to} \quad \boldsymbol{X\Lambda} + \boldsymbol{s}^- = \theta^i \boldsymbol{x}^i$$

$$\boldsymbol{Y\Lambda} - \boldsymbol{s}^+ = \boldsymbol{y}^i$$

$$\boldsymbol{s}^-, \boldsymbol{s}^+, \boldsymbol{\Lambda} \geq 0$$

As a result, DMU^i is efficient if $\theta^i = 1$ and $\boldsymbol{s}^- = \boldsymbol{s}^+ = 0$. However, DMU^i is weakly efficient if $\theta^i = 1$ and either $\boldsymbol{s}^- \neq 0$ or $\boldsymbol{s}^+ \neq 0$.

11.2.2.2 Variable Returns to Scale

CRS model assumes constant returns for all the DMU, whereas variable returns to scale (VRS) model does not. VRS models can contain constant returns to scale as well as increasing returns to scale and decreasing returns to scale. To that end, in VRS models convexity constraint on $\boldsymbol{\Lambda}$ is added to the previous CRS model.

The input-oriented VRS model in envelopment form is the following (Banker et al., 1984):

$$\text{minimize} \quad \theta - \varepsilon \mathbf{1}'\left(\boldsymbol{s}^- + \boldsymbol{s}^+\right)$$

$$\text{subject to} \quad \boldsymbol{X\Lambda} + \boldsymbol{s}^- = \theta\, \boldsymbol{x}^i$$

$$\boldsymbol{Y\Lambda} - \boldsymbol{s}^+ = \boldsymbol{y}^i$$

$$1'\boldsymbol{\Lambda} = 1$$

$$\boldsymbol{s}^-, \boldsymbol{s}^+, \boldsymbol{\Lambda} \geq 0$$

This VRS model is solved in two stage like the CRS model.

11.2.2.3 Window Analysis

By partitioning the time horizon into a number of windows, which contain a number of years, the efficiency score trends can be investigated. Same structured DEA models are run for each window only differing in the DMU and their respective inputs and outputs. This method was developed by Charnes et al. (1985).

Suppose that there are N decision making units (DMU) that are observed in T periods $(t=1,2,\ldots,T)$. Each DMU_t^i is the ith DMU $(i=1,\ldots,N)$ in period t with a s-dimensional input vector $\boldsymbol{x}_t^i=(x_{1t}^i,\ldots,x_{st}^i)'$ and a r-dimensional output vector $\boldsymbol{y}_t^i=(y_{1t}^i,\ldots,y_{rt}^i)'$. Also, suppose that a specific window starts at time k $(1\le k\le T)$ with the window width of w $(1\le w\le T-k)$. Then, the following input and output matrices are formed:

$$\boldsymbol{X}_{k_w}=\left[\boldsymbol{x}_k^1,\ldots,\boldsymbol{x}_k^N,\boldsymbol{x}_{k+1}^1,\ldots,\boldsymbol{x}_{k+1}^N,\ldots,\boldsymbol{x}_{k+w}^1,\ldots,\boldsymbol{x}_{k+w}^N\right]$$

$$\boldsymbol{Y}_{k_w}=\left[\boldsymbol{y}_k^1,\ldots,\boldsymbol{y}_k^N,\boldsymbol{y}_{k+1}^1,\ldots,\boldsymbol{y}_{k+1}^N,\ldots,\boldsymbol{y}_{k+w}^1,\ldots,\boldsymbol{y}_{k+w}^N\right]$$

Then, the input and output matrices of each window are fed separately to a specific DEA model. Here it is demonstrated for the case of input-oriented CRS model in envelopment form for a specific window k_w:

$$\begin{aligned}
\text{minimize} \quad & \theta-\varepsilon 1'\left(\boldsymbol{s}^-+\boldsymbol{s}^+\right)\\
\text{subject to} \quad & \boldsymbol{X}_{k_w}\boldsymbol{\Lambda}+\boldsymbol{s}^-=\theta\boldsymbol{x}_t^i\\
& \boldsymbol{Y}_{k_w}\boldsymbol{\Lambda}-\boldsymbol{s}^+=\boldsymbol{y}_t^i\\
& \boldsymbol{s}^-,\ \boldsymbol{s}^+,\ \boldsymbol{\Lambda}\ge 0
\end{aligned}$$

This way, the efficiency score of DMU^i will be obtained for each period $t\in k_w$. Additionally, its score for each window is found by running the model again with the input and output matrices belonging to the other windows.

11.2.2.4 Malmquist Index

The Malmquist index was introduced by Malmquist (1953) as an indicator to analyze the consumption of inputs. Färe et al. (1994) developed a DEA-based Malmquist index that measures productivity change over time.

To compute the index, a specific DEA model is used for two single-period and two mixed-period measures (Zhu, 2014; Cook & Seiford, 2009). Here, for the demonstration purposes input oriented constant returns to scale in multiplier form is presented.

Suppose that there are N decision-making units (DMU) that are observed in T periods ($t=1,2,\ldots,T$). Each DMU_t^i is the ith DMU ($i=1,\ldots,N$) in period t with a s-dimensional input vector $\boldsymbol{x}_t^i=(x_{1t}^i,\ldots,x_{st}^i)'$ and a r-dimensional output vector $\boldsymbol{y}_t^i=(y_{1t}^i,\ldots,y_{rt}^i)'$. Then, the following input and output matrices are formed:

$$\boldsymbol{X}_t=\left[\boldsymbol{x}_t^1,\ldots,\boldsymbol{x}_t^N\right]$$

$$\boldsymbol{Y}_t=\left[\boldsymbol{y}_t^1,\ldots,\boldsymbol{y}_t^N\right]$$

Also, let $\theta_t^o(x_{t+1}^o,y_{t+1}^o)$ denote efficiency score of DMU_t^o at time t with the input and output measures of time $t+1$.

Then, the calculation for the two single-period measures $\theta_t^o(x_t^o,y_t^o)$ and $\theta_{t+1}^o(x_{t+1}^o,y_{t+1}^o)$ are as follows:

$$\begin{aligned}\theta_t^o\left(x_t^o,y_t^o\right)= \text{ minimize } \quad & \theta^o\\ \text{subject to } & \boldsymbol{X}_t\Lambda\leq\theta^o\boldsymbol{x}_t^o\\ & \boldsymbol{Y}_t\Lambda\geq\boldsymbol{y}_t^o\\ & \Lambda\geq 0\end{aligned}$$

$$\begin{aligned}\theta_{t+1}^o\left(x_{t+1}^o,y_{t+1}^o\right)= \text{ minimize } \quad & \theta^o\\ \text{subject to } & \boldsymbol{X}_{t+1}\Lambda\leq\theta^o\boldsymbol{x}_{t+1}^o\\ & \boldsymbol{Y}_{t+1}\Lambda\geq\boldsymbol{y}_{t+1}^o\\ & \Lambda\geq 0\end{aligned}$$

Additionally, the calculation for the mixed-period measures $\theta_{t+1}^{o}(x_t^o, y_t^o)$ and $\theta_t^o(x_{t+1}^o, y_{t+1}^o)$ are as follows:

$$\begin{aligned}\theta_{t+1}^{o}\left(x_t^o, y_t^o\right) = \text{minimize} \;\; & \theta^o \\ \text{subject to} \;\; & \boldsymbol{X}_{t+1}\boldsymbol{\Lambda} \le \theta^o \boldsymbol{x}_t^o \\ & \boldsymbol{Y}_{t+1}\boldsymbol{\Lambda} \ge \boldsymbol{y}_t^o \\ & \boldsymbol{\Lambda} \ge 0\end{aligned}$$

$$\begin{aligned}\theta_t^{o}\left(x_{t+1}^o, y_{t+1}^o\right) = \text{minimize} \;\; & \theta^o \\ \text{subject to} \;\; & \boldsymbol{X}_{t}\boldsymbol{\Lambda} \le \theta^o \boldsymbol{x}_{t+1}^o \\ & \boldsymbol{Y}_{t}\boldsymbol{\Lambda} \ge \boldsymbol{y}_{t+1}^o \\ & \boldsymbol{\Lambda} \ge 0\end{aligned}$$

Then, the Malmquist productivity index is found as follows:

$$M^o = \left[\frac{\theta_t^o\left(x_t^o, y_t^o\right)}{\theta_t^o\left(x_{t+1}^o, y_{t+1}^o\right)} \frac{\theta_{t+1}^o\left(x_t^o, y_t^o\right)}{\theta_{t+1}^o\left(x_{t+1}^o, y_{t+1}^o\right)}\right]^{1/2}$$

It measures the change in productivity between periods t and $t+1$. If $M^o > 1$, then productivity declines. If $M^o = 1$, then productivity remains unchanged. If $M^o < 1$, then productivity improves.

The following modification of M^o allows the measurement of technical efficiency change and the frontier movement in terms of a specific DMU^o:

$$\overline{M}^o = \frac{\theta_t^o\left(x_t^o, y_t^o\right)}{\theta_{t+1}^o\left(x_{t+1}^o, y_{t+1}^o\right)} \left[\frac{\theta_{t+1}^o\left(x_{t+1}^o, y_{t+1}^o\right)}{\theta_t^o\left(x_{t+1}^o, y_{t+1}^o\right)} \frac{\theta_{t+1}^o\left(x_t^o, y_t^o\right)}{\theta_t^o\left(x_t^o, y_t^o\right)}\right]^{\frac{1}{2}}$$

The first term on right-hand side gives the magnitude of technical efficiency change (TEC) between periods t and $t+1$ as such $TEC^o = \frac{\theta_t^o(x_t^o, y_t^o)}{\theta_{t+1}^o(x_{t+1}^o, y_{t+1}^o)}$. $TEC^o > 1$ means improvement, $TEC^o = 1$ describes no change, whereas $TEC^o < 1$ indicates decline. The rest measures the frontier shift between periods t and $t+1$, namely $FS^o = \left[\frac{\theta_{t+1}^o(x_{t+1}^o, y_{t+1}^o)}{\theta_t^o(x_{t+1}^o, y_{t+1}^o)} \frac{\theta_{t+1}^o(x_t^o, y_t^o)}{\theta_t^o(x_t^o, y_t^o)}\right]^{\frac{1}{2}}$. For $FS^o > 1$ there is progress in frontier technology from period t to $t+1$. $FS^o = 1$ indicates status quo, whereas $FS^o < 1$ shows regress in frontier technology.

11.3 METHODOLOGY

In this study, DEA and cluster analysis are used for investigating relationship between competitiveness and human development levels of countries. Initially, the efficiencies of countries are calculated using DEA. Then the progress of countries through a time horizon are analyzed using a cluster analysis approach.

11.3.1 DEA Models

In this study, DEA is used to compute the investigated countries' efficiency scores with respect to their conversion of human development to competitiveness and competitiveness to human development. For this purpose, two DEA structures are designed:

DEA-1-HDItoGCI: INPUT: dimensions of HDI, OUTPUT: subindexes of GCI,

DEA-2-GCItoHDI: INPUT: subindexes of GCI, OUTPUT: dimensions of HDI.

Countries operate at different scales, that's why a BCC model is preferred. The model is output-oriented in each DEA structure as output enhancement is the goal. On the other hand, window analysis is used to investigate the temporal changes.

Suppose that there are N decision-making units (DMU) that are observed in T periods ($t=1,2,\ldots, T$). In our context, DMU are countries. Each DMU_t^i is the ith DMU ($i=1,\ldots, N$) in period t with a s-dimensional input vector $\boldsymbol{x}_t^i=(x_{1t}^i,\ldots,x_{st}^i)'$ and a r-dimensional output vector $\boldsymbol{y}_t^i=(y_{1t}^i,\ldots,y_{rt}^i)'$. Also, suppose that a specific window starts at time k ($1\leq k\leq T$) with the window width of w ($1\leq w\leq T-k$). Then, the following input and output matrices are formed:

$$\boldsymbol{X}_{k_w}=\left[\boldsymbol{x}_k^1,\ldots,\boldsymbol{x}_k^N,\boldsymbol{x}_{k+1}^1,\ldots,\boldsymbol{x}_{k+1}^N,\ldots,\boldsymbol{x}_{k+w}^1,\ldots,\boldsymbol{x}_{k+w}^N\right]$$

$$\boldsymbol{Y}_{k_w}=\left[\boldsymbol{y}_k^1,\ldots,\boldsymbol{y}_k^N,\boldsymbol{y}_{k+1}^1,\ldots,\boldsymbol{y}_{k+1}^N,\ldots,\boldsymbol{y}_{k+w}^1,\ldots,\boldsymbol{y}_{k+w}^N\right]$$

Then, the output-oriented BCC model in envelopment form for a window starting at k with width w is as follows (Zhu, 2014; Guccio et al., 2017):

$$\begin{aligned} \text{maximize} \quad & \Phi + \varepsilon 1'(s^- + s^+) \\ \text{subject to} \quad & X_{k_w}\Lambda + s^- = x_t^i \\ & Y_{k_w}\Lambda - s^+ = \Phi y_t^i \\ & 1'\Lambda = 1 \\ & s^-, s^+, \ \Lambda \geq 0 \end{aligned}$$

To include delays in the conversions, a 3-year time lag was considered in the models. As such input data for both DEA structures are between 2007 and 2014 whereas the output data's range is the years 2010–2017. The investigated time horizon h is 8 years. Number of windows, m, and the window width, w, is acquired as follows (Cooper et al., 2000):

$$m = h - w + 1$$

$$w = \begin{cases} 0.5 \times (h+1) & \text{, when } h \text{ is odd} \\ 0.5 \times (h+1) \mp 0.5 & \text{, when } h \text{ is even} \end{cases}$$

Therefore, in our case, because $h = 8$, number of windows is $m = 5$ and the window width is taken as $w = 4$ years. This is valid for both DEA structures.

DMU's yearly efficiency scores are calculated by taking the average of a specific year's efficiency scores found in all windows including that year. A sample result for the window analyzes of Turkey is given in Table 11.1.

A row contains the country's efficiency scores for each year belonging to the specific window. A row average indicates country's performance in a specific window, whereas the column average shows the country's performance in a specific year.

TABLE 11.1 Window Analyses of Turkey for DEA-1 Scores

	2010	2011	2012	2013	2014	2015	2016	2017
W1	0.895	0.905	0.932	0.909				
W2		0.901	0.918	0.891	0.887			
W3			0.918	0.891	0.887	0.864		
W4				0.900	0.892	0.869	0.860	
W5					0.897	0.869	0.856	0.865
Mean/Year	0.895	0.903	0.923	0.898	0.891	0.867	0.858	0.865

If the number of DMU are less than the number of inputs and outputs, DEA's discrimination performance suffers (Cook et al., 2014). Our study does not experience this problem as we consider a total of 6 inputs and outputs while each window contains 224 DMU. Additionally, Dyson et al. (2001) states using index data might be problematic for DEA applications, but it is acceptable if all the data are index type. Our case is admissible in this context because our data consists of only indexes.

11.3.2 Cluster Analysis

Cluster analysis is a multivariate analysis for organizing a set of observations or objects into groups called *clusters*. The observations within each group are close to each other whereas the clusters are dissimilar. In this particular study, the cluster analysis is used to classify the countries into groups depending on their human development, competitiveness, and DEA scores. By this way, it will be possible the see the progress of the countries through the years. Besides, it will also help us to determine which of the DEA models (i.e., DEA-1-HDItoGCI or DEA-2-GCItoHDI) gives more accurate results.

The two main approaches in the cluster analysis are hierarchical cluster analysis and K-means cluster analysis. Hierarchical clustering is a sequential process. The observations are combined to each other according to a pairwise distance matrix. The result is presented in a figure called *dendrogram*, by which the number of clusters and the cluster of each observation are specified. As one of the nonhierarchical cluster analysis methods, in K-means method the number of clusters are set prior to the analysis. The aim of the method is to assign the observations to the cluster with nearest mean by minimizing the within-cluster sum of squares (i.e., variance) of the clusters.

In this study, our aim of using cluster analysis is to cluster the countries into several development levels and track their clusters in the time-horizon. The input of the analysis is the data of countries under investigation for the competitiveness subindices and human development dimensions as well as the scores of DEA models. For the initial year the number of clusters (the development levels) of the countries are not known. Therefore, a hierarchical cluster analysis is applied to find the number of clusters and cluster the countries. For the subsequent years, the clusters should depend on the clusters of the countries in the previous year. Therefore, K-means cluster analysis is applied where the clusters of previous year is considered. By this way, we develop the following methodology to cluster the countries for time periods.

STEPWISE CLUSTER ANALYSIS FOR TIME PERIODS

For the initial year ($t = 1$):
  Apply the hierarchical cluster analysis. Draw a dendrogram.
  Determine the number of the clusters based on the dendrogram.
  Find the clusters of the countries.
For the subsequent years ($t = 2,\ldots,T$)
  Get the clusters of countries in the previous year ($t - 1$).
  Find the centroid of the clusters using the data of year t.
  Apply K-means algorithm initializing from the cluster centroids.

11.4 EMPIRICAL RESULTS

To apply proposed DEA models and cluster analysis approach given in the previous section data provided from WEF and UNDP is utilized. For human development, HDI dimensions (1) life expectancy index, (2) education index, and (3) income index are used, whereas the GCI subindexes are (1) basic requirements, (2) efficiency enhancers, (3) innovation and sophistication factors are for competitiveness of countries.

To have comparable set of countries, the countries comprising the 90% of the globe's population or gross domestic product (GDP) are taken into consideration. A total of 68 countries are selected according to 2007 population and GDP (current USD) World Bank data. However, GCI data for 12 countries were missing. As a result, 56 countries were investigated. To give idea of the data used in the models, analyzed countries' population and GDP (current USD) data of 2007 and HDI, GCI data of 2017 are presented in Table 11.2.

11.4.1 DEA Results

Two DEA structures are analyzed in this study: DEA-1-HDItoGCI and DEA-2-GCItoHDI. Input data for both DEA structures are between 2007 and 2014 whereas the output data's range is the years 2010–2017. In the case of DEA-1-HDItoGCI the input data are HDI dimensions 2007–2014 and the output data are GCI subindexes 2010–2017, whereas for DEA-2-GCItoHDI the input data are GCI subindexes 2007–2014 and the output data are HDI dimensions 2010–2017 (Table 11.3).

The models are then solved using the CPLEX solver in the General Algebraic Modeling System (GAMS). The results of DEA-1-HDItoGCI and DEA-2-GCItoHDI for the countries are presented in Tables 11.4 and 11.5, respectively.

TABLE 11.2 Data of Analyzed Countries

			Competitiveness Subindexes (2017)			Human Development Dimensions (2017)		
Country	**GDP (2007) (in billion USD)**	**Population (2007) (in million)**	**(1)**	**(2)**	**(3)**	**(1)**	**(2)**	**(3)**
Algeria	134.98	34.30	4.398	3.675	3.129	0.866	0.664	0.744
Argentina	287.53	39.97	4.099	4.004	3.562	0.873	0.816	0.788
Australia	851.96	20.83	5.704	5.290	4.685	0.97	0.929	0.918
Austria	388.69	8.30	5.700	5.032	5.301	0.95	0.852	0.924
Bangladesh	79.61	147.14	4.108	3.650	3.276	0.812	0.508	0.545
Belgium	471.82	10.63	5.483	5.146	5.185	0.943	0.893	0.913
Brazil	1,397.08	191.03	4.081	4.273	3.664	0.857	0.686	0.744
Canada	1,464.98	32.89	5.716	5.517	4.824	0.962	0.899	0.917
China	3,552.18	1,317.89	5.322	4.882	4.325	0.868	0.644	0.76
Colombia	207.42	44.37	4.333	4.376	3.667	0.839	0.676	0.735
Denmark	319.42	5.46	5.900	5.256	5.276	0.937	0.92	0.932
Egypt	130.48	79.54	4.049	3.898	3.353	0.795	0.604	0.701
Ethiopia	19.71	81.00	4.047	3.387	3.355	0.706	0.327	0.43
Finland	255.38	5.29	5.984	5.299	5.476	0.946	0.905	0.909
France	2,657.21	64.02	5.539	5.102	5.068	0.965	0.84	0.902
Germany	3,439.95	82.27	5.971	5.533	5.648	0.941	0.94	0.927
Greece	318.50	11.05	4.584	4.049	3.601	0.945	0.838	0.832
India	1,201.11	1,179.68	4.677	4.475	4.290	0.751	0.556	0.627
Indonesia	432.22	232.99	4.984	4.518	4.288	0.759	0.622	0.708
Ireland	269.92	4.40	5.675	5.087	4.933	0.948	0.918	0.95
Italy	2,203.05	58.44	4.879	4.462	4.451	0.972	0.791	0.886
Japan	4,515.26	128.00	5.665	5.393	5.551	0.983	0.848	0.901
Kenya	31.96	38.09	3.902	4.087	4.096	0.728	0.551	0.512
Korea, Rep.	1,122.68	48.68	5.774	4.928	4.846	0.959	0.862	0.889
Madagascar	7.34	19.43	3.482	3.300	3.215	0.713	0.498	0.394
Malaysia	193.55	26.63	5.549	4.939	4.909	0.853	0.719	0.841
Mexico	1,052.70	111.84	4.589	4.431	3.842	0.882	0.678	0.775
Morocco	79.04	31.23	4.789	3.944	3.563	0.862	0.529	0.649
Mozambique	9.37	22.19	2.746	3.114	2.982	0.598	0.385	0.361
Nepal	10.33	26.21	4.364	3.560	3.075	0.779	0.502	0.484
Netherlands	839.42	16.38	6.241	5.455	5.621	0.954	0.906	0.932
Nigeria	166.45	146.42	2.929	3.906	3.266	0.521	0.483	0.598
Norway	401.08	4.71	6.024	5.285	5.188	0.959	0.915	0.985
Pakistan	152.39	160.33	3.684	3.653	3.593	0.717	0.411	0.6
Peru	102.17	28.29	4.446	4.224	3.330	0.85	0.689	0.721

(Continued)

TABLE 11.2 (*Continued*) Data of Analyzed Countries

			Competitiveness Subindexes (2017)			Human Development Dimensions (2017)		
Country	**GDP (2007) (in billion USD)**	**Population (2007) (in million)**	**(1)**	**(2)**	**(3)**	**(1)**	**(2)**	**(3)**
Philippines	149.36	89.29	4.596	4.266	3.724	0.758	0.661	0.682
Poland	429.06	38.12	4.988	4.652	3.755	0.889	0.866	0.841
Portugal	240.17	10.54	5.116	4.578	4.184	0.945	0.759	0.847
Romania	175.93	20.88	4.565	4.276	3.276	0.855	0.762	0.819
Russian Fed.	1,299.71	142.81	4.924	4.593	3.756	0.788	0.832	0.829
Saudi Arabia	415.96	25.25	5.279	4.685	4.116	0.842	0.787	0.938
South Africa	299.03	49.89	4.276	4.393	4.143	0.668	0.708	0.722
Spain	1,479.34	45.23	5.154	4.837	4.171	0.974	0.824	0.882
Sri Lanka	32.35	19.81	4.507	3.812	3.755	0.854	0.749	0.714
Sweden	487.82	9.15	6.000	5.298	5.566	0.963	0.904	0.932
Switzerland	479.91	7.55	6.387	5.647	5.857	0.977	0.897	0.96
Tanzania	21.50	41.92	3.875	3.465	3.453	0.712	0.441	0.495
Thailand	262.94	66.20	5.058	4.617	3.918	0.854	0.661	0.762
Turkey	675.77	69.60	4.754	4.398	3.654	0.862	0.689	0.833
Uganda	12.29	30.59	3.801	3.558	3.446	0.618	0.525	0.424
Ukraine	142.72	46.51	4.175	4.093	3.545	0.802	0.794	0.664
United Arab Emirates	257.92	6.04	6.022	5.226	4.932	0.883	0.738	0.985
United Kingdom	3,074.36	61.32	5.648	5.552	5.338	0.95	0.914	0.902
United States	14,477.64	301.23	5.543	6.010	5.797	0.916	0.903	0.953
Venezuela	230.36	27.69	3.142	3.423	2.790	0.842	0.741	0.705
Vietnam	77.41	85.89	4.523	4.241	3.487	0.869	0.626	0.615

TABLE 11.3 DEA Structures

DEA Models	Inputs (2007–2014)	Outputs (2010–2017)
DEA-1-HDItoGCI	Life expectancy index	Basic requirements
	Education index	Efficiency enhancers
	Income index	Innovation and sophistication factors
DEA-2-GCItoHDI	Basic requirements	Life expectancy index
	Efficiency enhancers	Education index
	Innovation and sophistication factors	Income index

According to the results obtained from DEA-1-HDItoGCI; China, Ethiopia, Finland, India, Japan, Kenya, Malaysia, Mozambique, Nepal, Netherlands, Nigeria, Pakistan, Saudi Arabia, South Africa, Sweden, Switzerland, Uganda, United Arab Emirates, United Kingdom, and United States are found to be efficient at least one year. Among these countries: Ethiopia, Mozambique, Nepal, Nigeria, Pakistan, and Uganda perform poor on both HDI and GCI, whereas Switzerland has high scores in both HDI and GCI. On the other hand, Kenya has low HDI scores, and Japan, Sweden, and USA have high GCI scores.

TABLE 11.4 DEA-1-HDItoGCI Results

Country	2010	2011	2012	2013	2014	2015	2016	2017	Range
Algeria	0.795[b]	0.813	0.774	0.781	0.849[a]	0.804	0.795	0.805	0.076
Argentina	0.769	0.774[a]	0.763[b]	0.736	0.739	0.743	0.747	0.769	0.038
Australia	0.961[a]	0.956	0.959	0.952	0.947[b]	0.954	0.957	0.949	0.014
Austria	0.934	0.929	0.941[a]	0.925	0.929	0.918[b]	0.939	0.934	0.024
Bangladesh	0.894	0.921[a]	0.884	0.884[b]	0.887	0.889	0.890	0.906	0.037
Belgium	0.923	0.943[a]	0.937	0.923	0.927	0.927	0.935	0.921[b]	0.022
Brazil	0.904	0.901	0.922[a]	0.893	0.907	0.858	0.839[b]	0.849	0.084
Canada	0.976	0.981	0.989[a]	0.975	0.974	0.984	0.972[b]	0.975	0.018
China	1[a]	1[a]	0.988	0.987[b]	0.995	0.998	0.999	1[a]	0.013
Colombia	0.848[b]	0.855	0.856	0.856	0.863	0.874	0.881[a]	0.869	0.033
Denmark	0.977	0.977[a]	0.952	0.928[b]	0.957	0.964	0.952	0.956	0.049
Egypt	0.848[a]	0.818	0.792	0.771	0.765[b]	0.775	0.780	0.828	0.083
Ethiopia	1[a]	1[a]	0.992	0.982[b]	0.995	1[a]	1[a]	1[a]	0.018
Finland	0.989	0.995	1[a]	0.998	0.989	0.981	0.974[b]	0.975	0.026
France	0.947[a]	0.944	0.936	0.928	0.931	0.933	0.939	0.927[b]	0.020
Germany	0.979	0.975[b]	0.987	0.990	0.981	0.989	0.989	0.991[a]	0.016
Greece	0.781	0.761	0.755[b]	0.764	0.790	0.791[a]	0.781	0.784	0.036
India	1[a]	1[a]	1[a]	0.995	0.959[b]	0.963	1[a]	1[a]	0.041
Indonesia	0.952	0.949[b]	0.952	0.980	0.985	0.972	0.961	0.994[a]	0.045
Ireland	0.865[b]	0.867	0.894	0.897	0.905	0.928	0.934[a]	0.924	0.068
Italy	0.824	0.830	0.835[a]	0.818	0.814[b]	0.820	0.825	0.830	0.021
Japan	0.999	1[a]	1[a]	0.995	1[a]	0.996	0.988	0.983[b]	0.017
Kenya	0.988	1[a]	0.984	0.988	1[a]	0.975[b]	0.993	1[a]	0.025
Korea, Rep	0.920[b]	0.954	0.959[a]	0.942	0.929	0.946	0.949	0.953	0.039
Madagascar	0.890	0.859[b]	0.899	0.945[a]	0.938	0.920	0.929	0.936	0.086

(Continued)

TABLE 11.4 (*Continued*) DEA-1-HDItoGCI Results

Country	2010	2011	2012	2013	2014	2015	2016	2017	Range
Malaysia	0.964[b]	0.988	0.982	0.973	0.993	1[a]	0.982	0.981	0.036
Mexico	0.847[b]	0.863	0.878	0.873	0.854	0.860	0.887[a]	0.887	0.040
Morocco	0.978	0.9996[a]	0.982	0.965	0.981	0.965	0.965	0.959[b]	0.041
Mozambique	1[a]	0.992	0.972	0.979	0.995	1[a]	0.953	0.942[b]	0.058
Nepal	0.820[b]	0.840	0.857	0.903	0.945	0.961	0.968	1[a]	0.180
Netherlands	0.962[b]	0.971	0.981	0.967	0.967	0.980	0.988	1[a]	0.038
Nigeria	1[a]	1[a]	1[a]	1[a]	0.991	0.974[b]	0.980	0.979	0.026
Norway	0.935[b]	0.948	0.957	0.960	0.965[a]	0.964	0.961	0.948	0.030
Pakistan	0.990	1[a]	0.990	0.979	0.973	0.945	0.934[b]	0.953	0.066
Peru	0.874	0.885	0.891[a]	0.884	0.879	0.873	0.875	0.867[b]	0.024
Philippines	0.857[b]	0.880	0.914	0.927	0.957[a]	0.952	0.942	0.936	0.100
Poland	0.895	0.890	0.898[a]	0.883[b]	0.887	0.889	0.886	0.884	0.015
Portugal	0.869	0.868	0.862	0.856[b]	0.881[a]	0.877	0.865	0.877	0.025
Romania	0.833	0.808[b]	0.811	0.817	0.846	0.852[a]	0.839	0.832	0.044
Russian Federation	0.865[b]	0.867	0.896	0.906	0.917[a]	0.910	0.891	0.915	0.052
Saudi Arabia	0.967	1[a]	0.987	0.973	0.961	0.958	0.901	0.898[b]	0.102
South Africa	1[a]	1[a]	1[a]	1[a]	0.992	0.985	0.996	0.965[b]	0.035
Spain	0.874	0.873	0.884	0.873	0.869[b]	0.877	0.893[a]	0.891	0.024
Sri Lanka	0.889	0.902[a]	0.882	0.881	0.870	0.875	0.874	0.846[b]	0.057
Sweden	0.987	1[a]	0.990	0.975	0.957[b]	0.962	0.980	0.973	0.043
Switzerland	0.991	1[a]	1[a]	0.991[b]	0.993	0.997	0.997	1[a]	0.009
Tanzania	0.937	0.942[a]	0.938	0.913	0.913	0.902[b]	0.920	0.923	0.040
Thailand	0.918	0.913[b]	0.914	0.915	0.937	0.937	0.943	0.949[a]	0.035
Turkey	0.895	0.903	0.923[a]	0.898	0.891	0.867	0.858[b]	0.865	0.065
Uganda	1.000	1[a]	0.999	0.978[b]	0.983	0.999	1[a]	1[a]	0.022
Ukraine	0.841[b]	0.855	0.895[a]	0.870	0.884	0.859	0.849	0.861	0.054
United Arab Emirates	0.984	0.970[b]	0.991	0.986	1[a]	0.995	0.990	0.989	0.030
United Kingdom	0.970[b]	0.992	1[a]	0.988	0.991	0.988	0.993	0.987	0.030
United States	1[a]	0.991	0.992	0.988	0.985[b]	0.988	0.989	1[a]	0.015
Venezuela	0.695	0.693	0.687	0.671	0.654	0.659	0.670	0.651[b]	0.045
Vietnam	0.959	0.937	0.922[b]	0.924	0.932	0.949	0.956	0.964	0.042

[a] Maximum row value.

[b] Minimum row value.

TABLE 11.5 DEA-2-GCItoHDI Results

Country	2010	2011	2012	2013	2014	2015	2016	2017	Range
Algeria	0.971[b]	1[a]	0.990	0.969	0.978	1[a]	0.994	0.982	0.031
Argentina	0.997	1[a]	0.998	0.997	0.987[b]	0.993	1[a]	0.999	0.013
Australia	0.992[b]	0.994	0.998	0.998	0.997	0.998	1[a]	1[a]	0.008
Austria	0.972[b]	0.975	0.976	0.979	0.978	0.978	0.980	0.982[a]	0.009
Bangladesh	0.961	0.977[a]	0.949	0.922	0.905[b]	0.916	0.915	0.915	0.073
Belgium	0.972	0.971[b]	0.975	0.977	0.973	0.977	0.979[a]	0.979	0.008
Brazil	0.964[a]	0.936	0.925	0.896	0.893	0.891[b]	0.896	0.900	0.073
Canada	0.978[b]	0.981	0.983	0.985	0.985	0.988	0.991	0.991[a]	0.013
China	0.893[a]	0.886	0.885[b]	0.887	0.887	0.890	0.890	0.891	0.008
Colombia	0.901[a]	0.899	0.896	0.883	0.880[b]	0.881	0.884	0.883	0.021
Denmark	0.969[b]	0.988	0.989	1[a]	0.992	0.989	0.993	0.992	0.031
Egypt	0.887	0.879	0.857[b]	0.860	0.869	0.887	0.908	0.913[a]	0.056
Ethiopia	0.952[a]	0.842	0.846	0.815	0.791[b]	0.815	0.830	0.818	0.161
Finland	0.966[b]	0.969	0.971	0.977	0.975	0.975	0.976	0.977[a]	0.011
France	0.977[b]	0.980	0.983	0.985	0.984	0.986	0.987	0.989[a]	0.012
Germany	0.991[b]	0.994	0.999	0.999	0.999	1[a]	1[a]	1[a]	0.009
Greece	1[a]	0.988[b]	0.990	0.995	0.998	1[a]	1[a]	0.999	0.012
India	0.771[b]	0.775	0.783	0.776	0.784	0.785	0.792	0.795[a]	0.024
Indonesia	0.810[a]	0.795	0.790	0.790[b]	0.798	0.800	0.794	0.793	0.020
Ireland	1[a]	0.978[b]	0.990	0.994	1[a]	0.9996	1[a]	1[a]	0.022
Italy	0.999	1[a]	1[a]	0.9999	0.998[b]	0.998	0.9998	1[a]	0.002
Japan	0.994[b]	0.996	0.998	0.998	0.999	1[a]	1[a]	1[a]	0.006
Kenya	0.860	0.788[b]	0.895[a]	0.886	0.815	0.829	0.813	0.809	0.107
Korea, Rep	0.964[b]	0.968	0.977	0.984	0.983	0.982	0.985	0.988[a]	0.024
Madagascar	0.985	0.907	0.884	0.870	0.987[a]	0.945	0.869[b]	0.888	0.118
Malaysia	0.868[b]	0.873	0.879	0.882	0.880	0.884	0.887[a]	0.887	0.019
Mexico	0.919	0.920	0.922	0.927[a]	0.920	0.916[b]	0.919	0.925	0.011
Morocco	0.935	0.933	0.935[a]	0.930	0.922	0.918[b]	0.925	0.924	0.017
Mozambique	1[a]	0.941	0.905	0.788[b]	0.815	1[a]	0.864	0.877	0.212
Nepal	0.960	1[a]	1[a]	1[a]	0.945	0.917	0.898	0.881[b]	0.119
Netherlands	0.973[b]	0.981	0.983	0.982	0.982	0.983	0.984	0.986[a]	0.012
Nigeria	0.754	0.747[b]	0.806	1[a]	0.972	0.804	0.816	1[a]	0.253
Norway	0.998[b]	0.999	0.999	1[a]	1[a]	1[a]	1[a]	1[a]	0.002
Pakistan	0.805[b]	0.835	0.882	0.957	0.853	0.885	1[a]	0.981	0.195
Peru	0.945[a]	0.928	0.918	0.906	0.898[b]	0.901	0.902	0.907	0.047
Philippines	0.835	0.815	0.842[a]	0.830	0.806	0.796	0.795[b]	0.794	0.048
Poland	0.999	0.9996	0.985	0.992	0.965[b]	0.992	0.999	1[a]	0.035
Portugal	0.961[b]	0.964	0.968	0.972	0.972	0.974	0.975[a]	0.971	0.014

(Continued)

TABLE 11.5 (*Continued*) DEA-2-GCItoHDI Results

Country	2010	2011	2012	2013	2014	2015	2016	2017	Range
Romania	0.982[a]	0.968	0.958[b]	0.961	0.970	0.978	0.970	0.958	0.024
Russian Federation	0.971	0.968[b]	0.985	0.993	0.999	1[a]	0.991	0.974	0.032
Saudi Arabia	1[a]	0.997	0.999	0.990	0.976[b]	0.979	0.997	1[a]	0.024
South Africa	0.839[b]	0.841	0.853	0.848	0.850	0.852	0.857[a]	0.848	0.018
Spain	0.989[b]	0.990	0.994	1.000	0.998	0.997	0.999	1[a]	0.011
Sri Lanka	0.949[a]	0.946	0.937	0.909	0.902[b]	0.913	0.906	0.914	0.048
Sweden	0.977[b]	0.979	0.981	0.982	0.984	0.988	0.989	0.992[a]	0.015
Switzerland	0.998	0.997[b]	0.998	0.999	0.998	0.999	0.9996	1[a]	0.003
Tanzania	0.761[b]	0.795	0.778	0.794	0.798	0.800	0.819	0.824[a]	0.063
Thailand	0.862[b]	0.870	0.877	0.881	0.885	0.888[a]	0.886	0.887	0.026
Turkey	0.902[b]	0.923	0.929	0.939	0.940[a]	0.930	0.930	0.939	0.038
Uganda	0.950[a]	0.907	0.722	0.739	0.719	0.730	0.726	0.711[b]	0.239
Ukraine	0.982	0.970	0.991	0.993[a]	0.969	0.948[b]	0.956	0.949	0.046
United Arab Emirates	1[a]	1[a]	0.994[b]	0.999	1[a]	1[a]	1[a]	1[a]	0.006
United Kingdom	0.976	0.973[b]	0.978	0.988	0.992[a]	0.986	0.988	0.989	0.019
United States	0.989[b]	0.990	0.999	0.995	0.995	0.999	1[a]	1[a]	0.011
Venezuela	0.974[b]	0.996	1[a]	1[a]	1[a]	1[a]	1[a]	1[a]	0.026
Vietnam	0.945[a]	0.932	0.943	0.907[b]	0.914	0.922	0.924	0.929	0.038

[a] Maximum row value.
[b] Minimum row value.

DEA-2-GCItoHDI results indicate that Algeria, Argentina, Australia, Denmark, Germany, Greece, Ireland, Italy, Japan, Mozambique, Nepal, Nigeria, Norway, Pakistan, Poland, Russian Federation, Saudi Arabia, Spain, Switzerland, United Arab Emirates, United States, and Venezuela are efficient at least one year, whereas Mozambique and Nigeria have low GCI and HDI scores. Nepal and Venezuela perform poor on GCI scores. Australia and Norway have high HDI scores. Also, USA performs well on GCI score. Additionally, Denmark, Germany, and Switzerland have high GCI and HDI scores.

The number of efficient and inefficient countries are given in Table 11.6. The number of efficient countries varies between 2 and 12 in

TABLE 11.6 Number of Efficient and Inefficient Countries per Year

		2010	2011	2012	2013	2014	2015	2016	2017	Average
Number of efficient countries	DEA-1	7	12	7	2	3	3	3	9	5.75
	DEA-2	5	5	3	5	4	9	11	14	7
Number of inefficient countries	DEA-1	49	44	49	54	53	53	53	47	50.25
	DEA-2	51	51	53	51	52	47	45	42	49

DEA-1 and 3 and 14 in DEA-2. Because the number of efficient countries is not very high, the information provided by both models is rich and reliable.

According to the empirical results, efficient countries have different characteristics. Some have high GCI and HDI scores, and others have low scores. This result hinders the interpretability of our results based on only the GCI subindexes and HDI dimensions. That's why, cluster analysis was used to enhance the interpretability with the addition of country classes.

11.4.2 Cluster Analysis Results

The features DEA-1-HDItoGCI score, DEA-2-GCItoHDI score, HDI dimensions (Life expectancy index, Education index, Income index), and GCI subindexes (Basic requirements, Efficiency enhancers, Innovation and sophistication factors) are incorporated to the cluster analysis. Initially, hierarchical clustering was performed for the year 2010 by using Ward's method with squared Euclidean distance. The dendrogram shown in Figure 11.2 was obtained. The cut was made for four clusters. Furthermore, for the following years 2011–2017 K-means clustering was run with $K = 4$, where the centroids were fed from the previous cluster analysis results of the previous year to the algorithm. Thus, the K-means clustering was performed in order over the years. Table 11.7 presents the cluster analysis results. Also, Table 11.8 provides information about the cluster features.

According to the results in Table 11.7, 77% of countries (43 out of 56) remain in the same cluster in the investigated period. No drastic change is observed. Countries move either 1 up or 1 down in consecutive years. Some countries change their class once in the analysis period: Indonesia, Ireland, and Malaysia. Because their class improves in the years, these

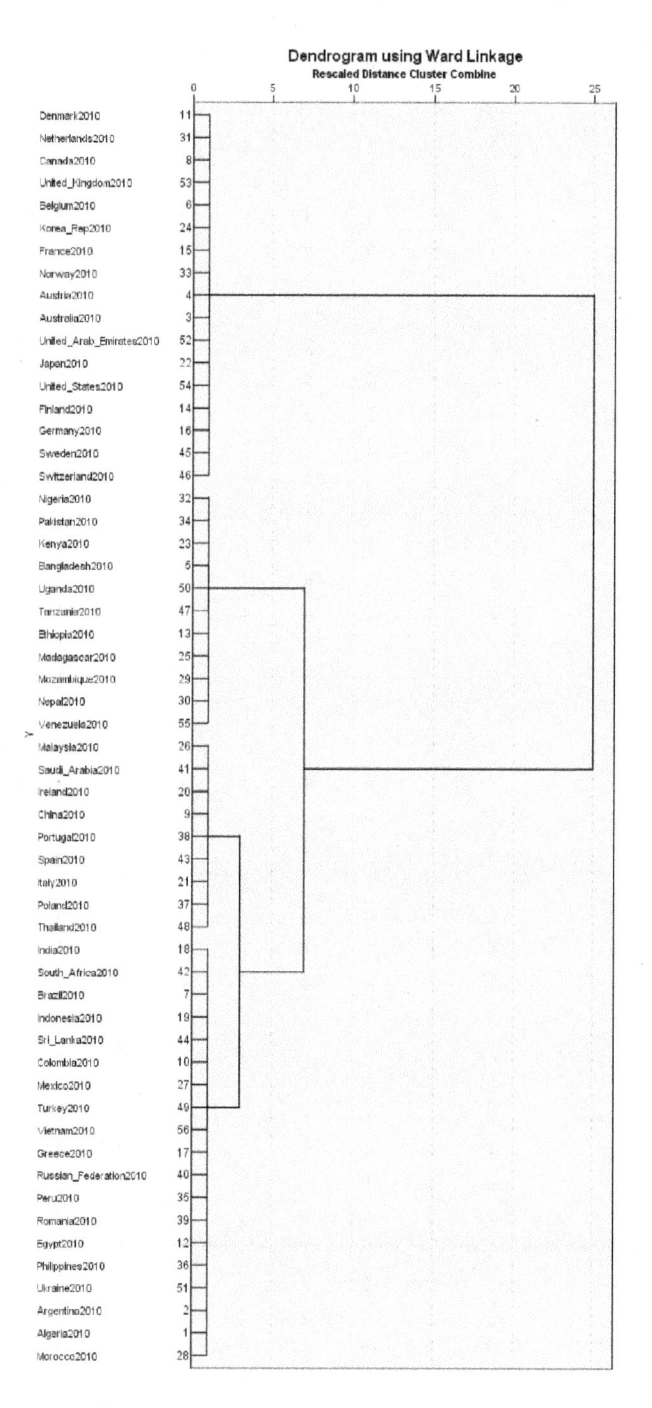

FIGURE 11.2 Dendrogram.

TABLE 11.7 Annual Cluster Membership

Country	2010	2011	2012	2013	2014	2015	2016	2017	Label
Algeria	2	1	1	1	2	1	1	2	Instable
Argentina	2	2	2	2	2	2	2	2	Stable
Australia	4	4	4	4	4	4	4	4	Stable
Austria	4	4	4	4	4	4	4	4	Stable
Bangladesh	1	1	1	1	1	1	1	1	Stable
Belgium	4	4	4	4	4	4	4	4	Stable
Brazil	2	2	3	3	2	2	2	2	Instable
Canada	4	4	4	4	4	4	4	4	Stable
China	3	3	3	3	3	3	3	3	Stable
Colombia	2	2	2	2	2	2	2	2	Stable
Denmark	4	4	4	4	4	4	4	4	Stable
Egypt	2	2	1	1	1	1	1	2	Instable
Ethiopia	1	1	1	1	1	1	1	1	Stable
Finland	4	4	4	4	4	4	4	4	Stable
France	4	4	4	4	4	4	4	4	Stable
Germany	4	4	4	4	4	4	4	4	Stable
Greece	2	2	2	2	2	2	2	2	Stable
India	2	2	3	3	2	2	3	3	Instable
Indonesia	2	2	3	3	3	3	3	3	On progress
Ireland	3	3	4	4	4	4	4	4	On progress
Italy	3	3	3	3	3	3	3	3	Stable
Japan	4	4	4	4	4	4	4	4	Stable
Kenya	1	2	1	2	2	2	2	2	Instable
Korea, Rep	4	4	4	4	4	4	4	4	Stable
Madagascar	1	1	1	1	1	1	1	1	Stable
Malaysia	3	4	4	4	4	4	4	4	On progress
Mexico	2	2	3	3	2	2	2	2	Instable
Morocco	2	2	2	2	2	2	2	2	Stable
Mozambique	1	1	1	1	1	1	1	1	Stable
Nepal	1	1	1	1	1	1	1	1	Stable
Netherlands	4	4	4	4	4	4	4	4	Stable
Nigeria	1	1	1	1	1	1	1	1	Stable
Norway	4	4	4	4	4	4	4	4	Stable
Pakistan	1	1	1	1	1	1	1	1	Stable
Peru	2	2	2	2	2	2	2	2	Stable
Philippines	2	2	2	2	2	2	2	2	Stable
Poland	3	3	3	3	3	3	3	3	Stable
Portugal	3	3	3	3	3	3	3	3	Stable

(Continued)

TABLE 11.7 (*Continued*) Annual Cluster Membership

Country	2010	2011	2012	2013	2014	2015	2016	2017	Label
Romania	2	2	2	2	2	2	2	2	Stable
Russian Federation	2	2	2	2	3	3	2	3	Instable
Saudi Arabia	3	4	4	4	3	3	3	3	Instable
South Africa	2	2	3	3	2	2	3	2	Instable
Spain	3	3	3	3	3	3	3	3	Stable
Sri Lanka	2	2	2	3	2	2	2	2	Instable
Sweden	4	4	4	4	4	4	4	4	Stable
Switzerland	4	4	4	4	4	4	4	4	Stable
Tanzania	1	1	1	1	1	1	1	1	Stable
Thailand	3	3	3	3	3	3	3	3	Stable
Turkey	2	2	3	3	3	2	2	2	Instable
Uganda	1	1	1	1	1	1	1	1	Stable
Ukraine	2	2	2	2	2	2	2	2	Stable
United Arab Emirates	4	4	4	4	4	4	4	4	Stable
United Kingdom	4	4	4	4	4	4	4	4	Stable
United States	4	4	4	4	4	4	4	4	Stable
Venezuela	1	1	1	1	1	1	1	1	Stable
Vietnam	2	2	2	2	2	2	2	2	Stable

1—Worst, 2—Poor, 3—Fair, 4—Best.

TABLE 11.8 Centroid Confidence Interval Values for All Years 2010–2017 ($\alpha = 0.05$)

Cluster	DEA-1	DEA-2	LEI	EI	II	BR	EE	ISF
1	[0.9, 0.92]	[0.88, 0.9]	[0.68, 0.71]	[0.47, 0.5]	[0.5, 0.53]	[3.58, 3.67]	[3.46, 3.5]	[3.09, 3.2]
2	[0.87, 0.89]	[0.91, 0.93]	[0.81, 0.83]	[0.67, 0.7]	[0.7, 0.72]	[4.37, 4.4]	[4.08, 4.15]	[3.5, 3.63]
3	[0.9, 0.92]	[0.92, 0.95]	[0.85, 0.88]	[0.7, 0.75]	[0.78, 0.82]	[4.83, 5.06]	[4.48, 4.59]	[3.98, 4.1]
4	[0.96, 0.97]	[0.98, 0.99]	[0.93, 0.94]	[0.86, 0.88]	[0.92, 0.92]	[5.69, 5.77]	[5.16, 5.26]	[5.12, 5.22]

countries can be named as countries on progress. Some others change their class more than once: Algeria, Brazil, Egypt, India, Kenya, Mexico, Russian Federation, Saudi Arabia, South Africa, Sri Lanka, and Turkey. These are instable countries that goes up and down in years.

Because the classes of the countries did not change dramatically, we can conclude that the cluster analysis results are robust.

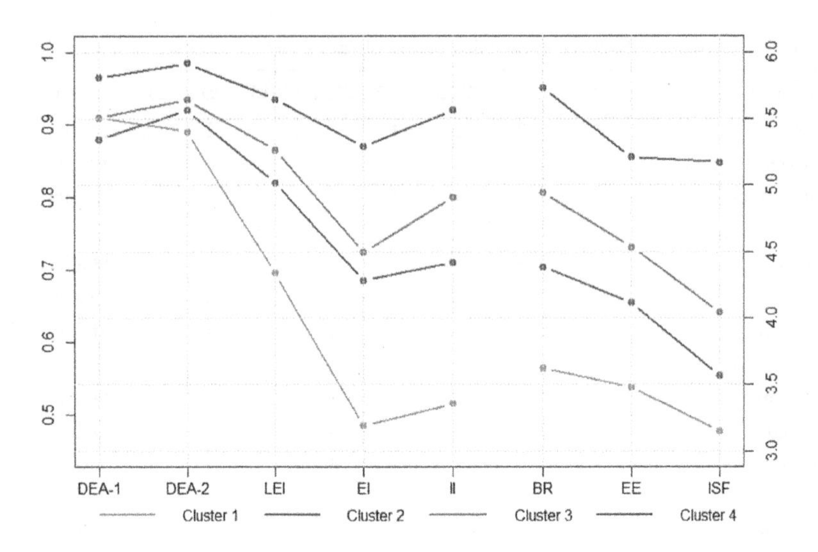

FIGURE 11.3 Cluster feature characteristics.

According to the centroids of the clusters all feature values except DEA-1-HDItoGCI scores are non-decreasing as the cluster membership ascends from worst to best or 1–4. Please see Table 11.8 and Figure 11.3 for the centroids of the clusters. Accordingly, cluster 1 performs the worst, whereas cluster 4 is the best cluster in terms of the features. Only exception to this pattern is DEA-1 scores. According to DEA-1 scores although the best cluster is Cluster 4, the other clusters are not ranked as in the other scores.

DEA-1 scores do not reflect the cluster memberships because according to this feature clusters 1 and 3 are same. Although cluster 4 is indeed shown to be the best by the DEA-1 scores, cluster 2 is worse than cluster 1, which is not the case based on the HDI dimensions and GCI subindexes. On the other hand, DEA-2 scores coincide with the classes. Only minor issue is that clusters 2 and 3 have overlapping values according to DEA-2 scores.

According to these results, because the cluster centroids are consistent with the other indices, we can conclude that DEA-2 model is more reliable than the DEA-1. Therefore, the direction of the relation between competitiveness and human development is from competitiveness to human development. The competitiveness level of a country in a certain year affects its human development with a time lag.

Finally, we ranked the countries according to their DEA-2 scores and cluster membership. The clusters are separated clearly from each other regarding the GCI and HDI features. As such, higher cluster number

TABLE 11.9 Ranking of the Countries Cluster Wise with DEA-2 Scores for the Year 2017

Country	Cluster	DEA-2	Rank	Country	Cluster	DEA-2	Rank
Australia	4	1	1–8	Indonesia	3	0.793	29
Germany	4	1	1–8	Argentina	2	0.999	30–31
Ireland	4	1	1–8	Greece	2	0.999	30–31
Japan	4	1	1–8	Algeria	2	0.982	32
Norway	4	1	1–8	Romania	2	0.958	33
Switzerland	4	1	1–8	Ukraine	2	0.949	34
United Arab Emirates	4	1	1–8	Turkey	2	0.939	35
United States	4	1	1–8	Vietnam	2	0.929	36
Denmark	4	0.992	9–10	Mexico	2	0.925	37
Sweden	4	0.992	9–10	Morocco	2	0.924	38
Canada	4	0.991	11	Sri Lanka	2	0.914	39
France	4	0.989	12–13	Egypt	2	0.913	40
United Kingdom	4	0.989	13–13	Peru	2	0.907	41
Korea, Rep	4	0.988	14	Brazil	2	0.900	42
Netherlands	4	0.986	15	Colombia	2	0.883	43
Austria	4	0.982	16	South Africa	2	0.848	44
Belgium	4	0.979	17	Kenya	2	0.809	45
Finland	4	0.977	18	Philippines	2	0.794	46
Malaysia	4	0.887	19	Nepal	1	0.881	47
Italy	3	1	20–23	Uganda	1	0.711	48
Poland	3	1	20–23	Nigeria	1	1	49–50
Saudi Arabia	3	1	20–23	Venezuela	1	1	49–50
Spain	3	1	20–23	Pakistan	1	0.981	51
Russian Federation	3	0.974	24	Bangladesh	1	0.915	52
Portugal	3	0.971	25	Madagascar	1	0.888	53
China	3	0.891	26	Mozambique	1	0.877	54
Thailand	3	0.887	27	Tanzania	1	0.824	55
India	3	0.795	28	Ethiopia	1	0.818	56

indicates higher ranking in these features. Within a cluster, the countries are ranked based on their DEA-2 scores. The ranking of the countries for year 2017 is given in Table 11.9.

11.5 CONCLUSIONS

In this study we analyzed the relationship between human development and competitiveness. Data of 56 selected countries from subindexes of GCI and dimensions of HDI for the years 2010–2017 are used in DEA models and cluster analysis. We developed two DEA models to structure mutual effects. In DEA-1, HDI dimensions are considered inputs and GCI subindexes are set as outputs, and vice versa in DEA-2. To see the effects with delays DEA models with time windows are applied considering a 3-year time-lag. Further, a stepwise cluster analysis for time periods is used to enrich the interpretation of the results of the DEA models. The countries are clustered according to their DEA-1, DEA-2 results, and scores of HDI dimensions and GCI subindexes. The results are found to be robust and consistent. Finally, the countries are ranked based on their clusters and DEA-2 scores.

According to the results, competitiveness affects the human development in time horizon with a delay. This result is consistent with some studies in the literature. For instance, Waheeduzzaman (2002) proved that competitiveness has positive impact on human development. According to Ülengin et al. (2011) if the competitiveness of a country is managed properly, then the level of human well-being is expected to improve. Aiginger (2006) considered the human development as an out of the competitiveness and presents a model for measuring competitiveness via its output, welfare.

The basic implication of this result is that the countries needing a development in human well-being can focus on the improvements on their competitiveness. Developments in the basic requirements (such as institutions, infrastructure, macroeconomic environment, etc.), efficiency enhancers (such as higher education and training, labor market efficiency, etc.), and innovation and sophistication factors (business sophistication and innovation) will lead to a longer and healthy life, higher knowledgeable citizens, and higher standard of living. This perspective can also support the belief that the goal of competitiveness is to enhance the prosperity of the citizens.

ACKNOWLEDGMENTS

This work is supported by İstanbul Technical University, BAP (Project ID: SGA-2017-40636).

REFERENCES

Aiginger, K. (1998). A framework for evaluating the dynamic competitiveness of countries. *Structural Change and Economic Dynamics, 9*, 159–188.

Aiginger, K. (2006). Competitiveness: From a dangerous obsession to a welfare creating ability with positive externalities. *Journal of Industry, Competition and Trade, 6*, 161–177.

Anand, S. & Sen, A. (1992). Human development index: Methodology and measurement. *Human Development Report Office*. New York: UNDP.

Banker, R. D., Charnes, A., & Cooper, W. W. (1984). Some models for estimating technical and scale inefficiencies in data envelopment analysis. *Management Science, 30*, 1078–1092.

Bérenger, V., & Verdier-Chouchane, A. (2007). Multidimensional measures of well-being: Standard of living and quality of life across countries. *World Development, 35*(7), 1259–1276.

Bouyssou, D. (1999). Using DEA as a tool for MCDM: Some remarks. *Journal of the Operational Research Society, 50*(9), 974–978.

Bucher, S. (2018). The Global Competitiveness Index as an indicator of sustainable development. *Herald of Russian Academy of Sciences, 88*(44), 44–57. doi:10.1134/S1019331618010082.

Buscema, M., Sacco, P., & Ferillli, G. (2016). Multidimensional similarities at a global scale: An approach to mapping open society orientations. *Social Indicators Research, 128*, 1239–1258.

Charnes, A., Clarke, C., Cooper, W. W., & Golany, B. (1985). A development study of DEA in measuring the effect of maintenance units in the U.S. Air Force. *Annals of Operations Research, 2*, 95–112.

Charnes, A., Cooper, W., & Rhodes, E. (1978). Measuring the efficiency of decision making units. *European Journal of Operational Research, 2*, 429–444.

Ciko, D. (2015). Albania vs. Balkan's countries as comparative analysis of the human development index. *Mediterranean Journal of Social Sciences, 6*(4), 311–318.

Cook, W., & Seiford, L. (2009). Data envelopment analysis (DEA): Thirty years on. *European Journal of Operational Research, 192*(1), 1–17.

Cook, W., Tone, K., & Zhu, J. (2014). Data envelopment analysis: Prior to choosing a model. *Omega, 44*(2014), 1–4.

Cooper, W., Seiford, L., & Tone, K. (2000). *Data Envelopment Analysis: A Comprehensive Text with Models, Applications, References and DEA-Solver Software*. New York: Springer.

Ditkun, S., Klafke, R., Ahrens, R., Kovaleski, J., & Canabarro, N. (2014). The ranking of Brazil in global competitiveness: A study of its evolution in the period 2003–2013. *Espacios, 35*(10), 8. http://www.revistaespacios.com/a14v35n10/14351008.html

Dudas, S. (2014). The impact of the global economic crisis of 2008/2009 on the national competitiveness of Central and Eastern European countries. In M. Dolinsky, & V. Kunova içinde (Eds.), *Current Issues of Science and Research in the Global World* (pp. 99–106). London, UK: CRC Press.

Dyson, R. G., Allen, R., Camanho, A. S., Podinovski, V. V., Sarrico, C. S., & Shale, E. A. (2001). Pitfalls and protocols in DEA. *European Journal of Operational Research, 132*, 245–259.

Edewor, P. A. (2014). A conceptual exploration of the human development paradigm. *Mediterranean Journal of Social Sciences, 5*(6), 381–388.

Emrouznejad, A. & Yang, G. (2018). A survey and analysis of the first 40 years of scholarly literature in DEA: 1978–2016. *Socio-Economic Planning Sciences, 61*, 4–8.

Färe, R. S., Grosskopf, S., & Lovell, C. A. K. (1994). *Production Frontiers.* Cambridge, UK: Cambridge University Press.

Guccio, C., Martorana, M. F., & Mazza, I. (2017). The efficiency change of Italian public universities in the new millennium: A non-parametric analysis, *Tertiary Education and Management, 23*(3), 222–236, doi:10.1080/13583883.2017.1329451.

Jiyad, A. M. (1998). Human development paradigm under globalization environment. *Nordic Conference on Middle Eastern Studies: The Middle East in Globalizing World.* Oslo, Norway.

Liu, J., Lu, L., & Lu, W. (2016). Research fronts in data envelopment analysis. *Omega, 58*, 33–45.

Liu, J., Lu, L., Lu, W., & Lin, B. (2013). A survey of DEA applications. *Omega, 41*, 893–902.

Lonska, J., & Boronenko, V. (2015). Rethinking competitiveness and human development. *Procedia Economics and Finance, 23*, 1030–1036.

Malmquist, S., 1953. Index numbers and indifference surfaces. *Trabajos de Estatistica, 4*, 209–242.

Onyusheva, I. (2015). Human capital in conditions of global competitiveness: The case of Kazakhstan. *Proceedings of the 12th International Conference on Intellectual Capital, Knowledge Management and Organisational Learning*, (pp. 191–196). Bangkok, Thailand.

Perez-Moreno, S., Rodriguez, B., & Luque, M. (2016). Assessing global competitiveness under multi-criteria perspective. *Economic Modelling, 53*, 398–408.

Porter, M. (1990). *The Competitive Advantage of Nations.* Cambridge, UK: Harvard Business Review.

Sánchez, J. J. V. (2018). Malmquist index with time series to data envelopment analysis. In V. Salomon (Ed.), *Multi-criteria Methods and Techniques Applied to Supply Chain Management* (pp. 111–130). London, UK: IntechOpen.

Seers, D. (1972). What are we trying to measure? *The Journal of Development Studies, 8*(3), 21–36.

Sen, A. (1985). *Commodities and Capabilities.* Amsterdam, the Netherlands: North Holland.

Sen, A. (1992). *Inequality Re-examined.* Oxford, UK: Clarendon Press.

Shkiotov, S. (2013). M. Porter's national competitiveness model verification: Correlation between the level of national competitiveness, labor productivity and the quality of life. *World Applied Sciences Journal, 25*(4), 684–689.

Skorvagova, S., & Drienikova, K. (2016). Socioeconomic aspects of the EU competitiveness. *Actual Problems of Economics, 6*(180), 55–66.

Thore, S., & Tarverdyan, R. (2016). The sustainable competitiveness of nations. *Technological Forecasting & Social Change 106*, 108–114.

Tridico, P., & Meloni, W. P. (2018). Economic growth, welfare models and inequality in the context of globalisation. *The Economic and Labour Relations Review, 29*(1), 118–139. doi:10.1177/1035304618758941.

UNDP (1990). Concept and measurement of human development. *Human Development Report*. New York: UNDP.

Ülengin, F., Kabak, Ö., Önsel, Ş., & Aktaş, E. (2009). Assessment of implication of competitiveness on human development of countries through data envelopment analysis and cluster analysis. *Financial Modeling Applications and Data Envelopment Applications, 13*, 199–226.

Ülengin, F., Kabak, Ö., Önsel, Ş., Aktaş, E., & Parker, B. (2011). The competitiveness of nations and implications for human development. *Socio-Economic Planning Sciences, 45*, 16–27.

Waheeduzzaman, A. (2002). Competitiveness, human development and inequality: A cross-national comparative inquiry. *Competitiveness Review, 12*(2), 13–29.

World Economic Forum. (2011). *The Global Competitiveness Report 2010–20211.* Versoix, Switzerland: SRO-Kundig.

Zhu, J. (2014) *Quantitative Models for Performance Evaluation and Benchmarking: Data Envelopment Analysis with Spreadsheets.* Heidelberg, Germany: Springer.

Index

Note: Page numbers in italic and bold refer to figures and tables, respectively.